# Thengiz Beridze

# Satellite DNA

With 78 Figures

Springer-Verlag
Berlin Heidelberg New York Tokyo

Professor Dr. Thengiz Beridze
Institute of Plant Biochemistry
Academy of Sciences of the Georgian SSR
SU-Tbilisi 380031, UdSSR

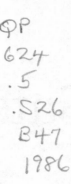

Revised and enlarged translation of the Russian edition (Nauka, Moscow, 1982).

ISBN 3-540-15876-6 Springer-Verlag Berlin Heidelberg New York Tokyo
ISBN 0-387-15876-6 Springer-Verlag New York Heidelberg Berlin Tokyo

Library of Congress cataloging in Publication Data.
Beridze, Thengiz, 1939–. Satellite DNA.
Translation of: Satellitnye DNK.
Bibliography: p.
Includes index.
1. Satellite DNA. I. Title.
QP624.5.S26B4713    1986    574.87'3282    86-1861
ISBN 0-387-15876-6 (U.S.)

Typesetting: With a system of the Springer Produktions-Gesellschaft, Berlin.
Dataconversion: Daten- und Lichtsatz-Service, Würzburg.
Printing: Color-Druck, Berlin. Bookbinding: Lüderitz & Bauer-GmbH, Berlin.
2131/3020-543210

# Preface

The organization of genomes in higher organisms has been studied extensively in recent years. With current achievements in gene engineering, it seems quite realistic that they will be specifically modified in the nearest future to produce new, economically valuable forms of animals and plants. The success of these experiments will depend greatly on the level of our knowledge of the structural features of plant and animal DNAs.

Comparative studies of DNA from different organisms began with discovery of its genetic significance in the late 1950's. A few years later it was found that nuclear DNA, the main storage of genetic information, can consist of several fractions differing in some physical and chemical properties. Along with the "major" DNA, bearing the main load during the genotype functioning, the so-called satellite DNAs were discovered.

T. G. Beridze, the author of this book, is one of the pioneers in the study of these extraordinary DNAs. The results of his experiments with plant satellite DNAs have essentially influenced the formation of our current ideas on their structure and properties.

There is no need to recapitulate the contents of the book in the Preface. It is sufficient to say that the book provides a comprehensive description of the physical and chemical properties of various satellite DNAs and discusses in detail the current viewpoints on their functional role. Although satellite DNAs have been studied for over 20 years, at present neither their origin, nor their metabolic functions have been completely revealed and understood. In my opinion, the establishment of the structural and functional heterogeneity of satellite DNAs is the only clear conclusion that can be drawn from all the preceding studies.

The analysis of the presented material and the hypotheses proposed by the author should stimulate more extensive studies of this unusual component of the genomes of higher organisms. Today there are grounds to believe that satellite DNAs may appear (at least in some cases) as a result of misfunctioning of the gene amplification mechanism. Thus, studies of the formation pathways of satellite DNAs can shed light on the molecular aspects of the process which is extremely important for some stages of plant and animal ontogenesis, for the evolutionary transformation of genotypes. Of no less interest is the development and experimental testing of hypotheses on the role of these DNAs in a number of important processes underlying the functioning of the cell genetic apparatus. As the author of the book has quite justly noted, the success of the work will depend on the effectiveness of cooperation of specialists in genetics, biochemistry, and cytology.

The main content of the book is the thoroughly arranged and well interpreted information on satellite DNAs. What makes the book more valuable is that the reader will get it "first-hand" from a scientist who has devoted many years to the investigation of satellite DNAs. I am sure that the information presented here will be welcomed and utilized by those interested in the study of the structure and organizational principles of genomes of higher organisms.

Moscow State University                                        Professor A. S. Antonov

# Contents

# Abbreviations

| | | |
|---|---|---|
| stDNA | – | Satellite DNA |
| nDNA | – | Nuclear DNA |
| mtDNA | – | Mitochondrial DNA |
| cpDNA | – | Chloroplast DNA |
| hrDNA | – | Highly reiterated DNA |
| L-strand | – | Light strand of stDNA |
| H-strand | – | Heavy strand of stDNA |
| cRNA | – | Complementary RNA synthesized on a stDNA as a matrix |
| bp | – | Base pairs |
| kb | – | $10^3$ Base pairs |
| $m^5C$ | – | 5-Methylcytosine |
| $C_0t$ | – | Product of DNA concentration (moles of nucleotides per liter) and time of reassociation (seconds) |
| $r_f$ | – | Molar ratio of $Hg^{2+}$ or $Ag^+$ to DNA phosphate |
| $1 \times SSC$ | – | 0.15 M NaCl, 0.015 M Na citrate, pH 7.0 |
| $T_m$ | – | "Melting" point of DNA; a temperature at which the UV-light absorption reaches half of its maximum value |
| $\Delta T_m$ | – | Difference in $T_m$ between native and reassociated DNA |
| Kinetic complexity | – | Number of base pairs in an unrepeating DNA sequence |

The buoyant density values of DNA molecules indicated in the book correspond to the CsCl neutral density gradient, if not specially indicated (the *E.coli* DNA buoyant density is assumed to be $1.710 \, g/cm^3$).

Recognition Sites of Some Restriction Endonucleases Given in the Book [1]

| Enzyme | Recognition sequence |
|--------|---------------------|
| *Alu*I | AG↓CT |
| *Bam*HI | G↓GATCC |
| *Bsp*I | GGCC |
| *Bsu*RI | GG↓CC |
| *Eco*RI | G↓AATTC |
| *Eco*RI* | PuPuA↓TPyPy |
| *Eco*RII | ↓CC($^A_T$)GG |
| *Hae*III | GG↓CC |
| *Hha*I | GCG↓C |
| *Hind*III | A↓AGCTT |
| *Hinf*I | G↓ANTC |
| *Hpa*II | C↓CGG |
| *Hph*I | GGTGA – Cleaving occurs at eight nucleotides from the recognition site |
| *Kpn*I | GGTAC↓C |
| *Mbo*I | ↓GATC |
| *Mbo*II | GAAGA – Cleaving occurs at eight nucleotides from the recognition site |
| *Mnl*I | CCTC |
| *Sma*I | CCC↓GGG |
| *Taq*I | T↓CGA |
| *Sau*3AI | GATC |
| *Sau*96I | G↓GNCC |

[1] Cleavage sites are marked with arrows; unknown cleavage sites are not marked

# Introduction

The study of the structural organization of the eukaryotic genome is one of the central problems of present-day molecular biology. Apart from the unique and moderately repetitive fractions, the eukaryotic genome contains a fraction of satellite DNA consisting of highly repetitive tandemly arranged sequences. Their amount may be quite significant in many organisms, and can exceed even half of the genome.

Of the eukaryotic DNA fraction, the structure of satellite DNA is the one most thoroughly studied. At present, the primary structure of repeating units of satellite DNAs has been established in a number of organisms. But a precise knowledge of the chemical structure does not reveal any function of satellite DNAs. Satellite DNAs do not determine the structure of any protein. They are localized mainly in the constitutive heterochromatin, i.e., in the genetically inert and compact parts of chromosomes. The compact state of heterochromatin apparently depends on the presence of satellite DNAs. How does this proceed? What keeps the heterochromatin compact? What role do the structural features of satellite DNAs play in the process of chromatin compactization? None of the available data have so far shed light on the mechanism of chromatin compactization or have allowed to disclose any features common for satellite DNAs on the whole.

It is quite clear that any model of the structural and functional organization of an eukaryotic genome will be an unsatisfactory one if it fails to define exactly the site of highly repetitive DNAs and, in particular, of satellite DNAs. Together with this, determination of the structural features and elucidation of the functional role of satellite DNAs are important not only to clarify the eukaryotic genome organization, but also to give an insight into the evolution of eukaryotes.

Some reviews devoted to separate aspects of satellite DNA problems have been published (John and Miklos 1979; Brutlag 1980; Singer 1982). This book is a first attempt to summarize all the available experimental data accumulated over the past 20 years since the discovery of satellite DNAs and to consider the theoretical conceptions on the origin, evolution, and functional role of satellite DNAs.

The title itself, "Satellite DNA", indicates the versatility of satellite DNAs in the length of repeats, in the GC content, and amount contained in the genome.

The data in the book do not include all the satellite DNAs that are known at present. The presented data refer only to satellite DNAs most thoroughly studied in the structural aspect. This information is necessary for discussing more general problems concerning the origin of satellite DNAs, their evolution, the structure of satellite DNA-containing chromatin, and the functional role of satellite DNAs.

The author's viewpoint on the most essential questions is presented in the chapters devoted to the general problems of the origin and functional role of satellite DNAs. Problems requiring solution and which will be apparently the subject of detailed research in the nearest future are outlined.

One purpose of this book is also to present the evolution of the term "satellite DNA" in retrospect. Although only about 20 years have elapsed since the discovery of satellite DNAs, such an exposition of the material permits to follow the formation of up-to-date concepts in this field of research with greater clarity. The term satellite DNA itself has undergone a significant evolution since 1961 when it was used for the first time, and at the present is used in a somewhat different sense. The chapter devoted to the evolution of the term satellite DNA seems to reflect fully all shades of the different definitions of this fraction.

Just as all the other branches of molecular biology, the state of the art considered here is developing intensively and may soon be outdated. However, there are many key points in the problem of satellite DNAs which have been fully formulated in the book, and this will considerably prolong its life.

In conclusion, I would like to express my deep gratitude to Prof. S. V. Durmishidze of the Academy of Sciences of the Georgian SSR for his valuable advice and friendly assistance during preparation of the manuscript. I also thank my colleagues for reading the manuscript and helpful remarks.

# 1 Detection of Satellite DNAs

Detection of satellite DNAs was stimulated by the development of density gradient ultracentrifugation methods. Equilibrium ultracentrifugation in the density gradient suggested by Meselson and co-workers (Meselson et al. 1957) is one of the most valuable methods of DNA fractionation and characterization. The method is based on the equilibrium distribution of the macromolecular material in the density gradient by sedimentation of a low molecular substance solution in a constant centrifugal field. At equilibrium, a stable concentration gradient of the low molecular substance is formed under the effect of two opposite actions: sedimentation and diffusion and, as a result, a gradual increase of density is observed towards the direction of the centrifugal force. The initial concentration of the low molecular substance, the centrifugal field, and the height of the liquid column can be so selected that at equilibrium the macromolecules will be pooled into the density gradient zone. The centrifugal field directs the macromolecules to a region where the sum of the forces acting on a given molecule is equal to zero. This tendency is opposed by the Brownian motion, and as soon as equilibrium is reached, the macromolecules form a zone in the density gradient, with a width inversely proportional to the molecular weight (Fig. 1).

The value of buoyant density is used to characterize the distribution of macromolecules in the gradient. It is equal to the density of the position in the gradient at which the sum of the forces acting on the macromolecules is equal to zero (Vinograd and Hearst 1962). The buoyant density of biopolymers varies within a wide range,

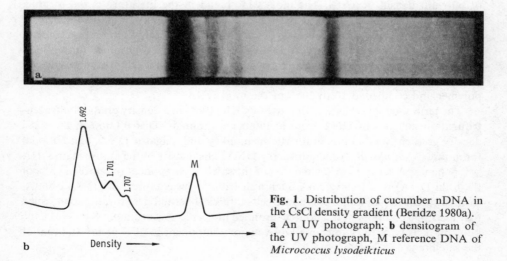

**Fig. 1**. Distribution of cucumber nDNA in the CsCl density gradient (Beridze 1980a). **a** An UV photograph; **b** densitogram of the UV photograph, M reference DNA of *Micrococcus lysodeikticus*

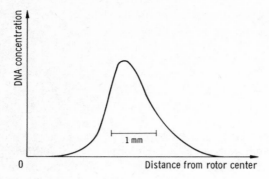

**Fig. 2**. Distribution of calf thymus DNA in the CsCl density gradient (Meselson et al. 1957)

e. g., in the CsCl density gradient the buoyant density of proteins is 1.3 g/cm³, and 1.7 g/cm³ for DNA, while RNA in a saturated solution with a density of 1.90 g/cm³ sediments to the bottom of the gradient.

The first eukaryotic DNA analyzed in this way was calf thymus DNA. It revealed an asymmetrical distribution in the CsCl density gradient indicating the heterogeneity of the material (Fig. 2). The phage T4 DNA was characterized by a Gaussian distribution in the given gradient.

Later, Schildkraut et al. (1962) confirmed the distribution asymmetry of calf thymus DNA and showed it to be due to the presence of a satellite component with a density of 1.713 g/cm³ (the buoyant density of the main DNA was 1.699 g/cm³). They also demonstrated in this study that the buoyant density of DNA in CsCl is directly proportional to the GC content. If a DNA preparation includes several fractions with different base composition, they will separate into several bands in the ultracentrifuge cell. Such a relation was found to be true also for denatured DNA, whose density exceeded that of native DNA by about 0.015 g/cm³. Several authors have suggested a formula to calculate the GC content in DNA molecules on the basis of buoyant density in CsCl. The most widespread in the literature is the formula $\varrho = 1.660 + 0.098$ (GC), derived by Schildkraut et al. (1962) from buoyant density data processing and chemical analysis of the GC content of DNA from 51 organisms. The formula is valid, provided the DNA does not contain any minor nucleotides.

About the same time Sueoka and Cheng (1961) detected a DNA satellite component in crabs of the genus *Cancer*. The component was similar to a synthetic dAT-polymer and comprised 10 to 30% of the total DNA.

The term satellite DNA was first used by Kit (1961) in a density gradient ultracentrifugation analysis of DNA from a number of animals. Distribution in the CsCl density gradient was unimodal for rhesus monkey and alligator DNA, and bimodal for mouse (*Mus musculus*) and guinea pig DNA. The density of the mouse main DNA component was 1.701 g/cm³, while that of the satellite component (comprising 8% of the total DNA) was 1.690 g/cm³. Such a distribution was typical of DNAs isolated from different tissues and from mouse cell cultures. Thermal denaturation confirmed that the 1.690 component is a double-stranded molecule. According to Kit (1961), the buoyant density of the guinea pig DNA major component is 1.697 g/cm³ and that of the satellite is 1.703 g/cm³.

According to Kit's definition, a satellite DNA is a minor fraction revealed as a separate band at equilibrium centrifugation in the CsCl analytical density gradient.

The nature of the detected satellite components remained obscure. Kit's data gave no indication on the intracellular localization of stDNAs. To explain the presence of these components, Schildkraut and co-workers (1962) suggested that the minor bands could be the result of the presence of symbiotic organisms, or that they could be a certain DNA fraction enriched by some minor base, such as 5-methylcytosine.

# 2 Evolution of the Term "Satellite DNA"

## 2.1 Intracellular Localization of stDNAs

Chun et al. (1963) published a paper which proved to be important for clarifying the nature of the DNA minor fractions. The work dealt with the character of distribution in the CsCl density gradient of two representatives of higher plant DNAs, spinach (Fig. 3) and beet. It was shown that the nDNA of these organisms is characterized by a unimodal distribution ($\varrho = 1.695\ \text{g/cm}^3$) in the CsCl density gradient. The total DNA revealed satellite components ($\varrho = 1.705$ and $1.719\ \text{g/cm}^3$) composing 10% of the genome. The proportion of these fractions increased in the DNA preparations isolated from chloroplasts. The conclusion followed that the 1.705 component is a cpDNA. With regards to the 1.719 component, the possibility of its localization in chloroplasts was not excluded (it was shown later that the 1.719 component resulted from bacterial contamination of the leaf surfaces).

These authors also found satellite components in the cells of photosynthesizing green algae *Chlamydomonas reinhardi* and *Chlorella ellipsoidea*. The buoyant density of the major component of *Chlamydomonas* and *Chlorella* was $1.723\ \text{g/cm}^3$ and

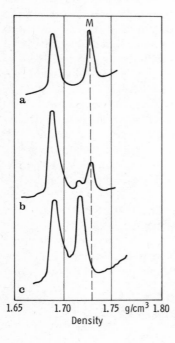

**Fig. 3**. Microdensitograms of spinach leaf DNA distribution in the CsCl density gradient (Chun et al. 1963). **a** Nuclear fraction; **b** total leaf DNA; **c** chloroplast fraction; M DNA of *M.lysodeikticus*

$1.716 \text{ g/cm}^3$, respectively, while that of the minor component was $1.695 \text{ g/cm}^3$, their proportion being about 1 %. According to the authors, the 1.695 component could be localized in chloroplasts. Later, using Chun's method, satellite components were detected in the cells of some other organisms and their localization was proved both in plastids and mitochondria. DNA was isolated from mitochondria of *Neurospora crassa* (Luck and Reich 1964), yeast (Tewari et al. 1965), *Euglena gracilis* (Edelman et al. 1966), and other organisms. The buoyant density of lower eukaryotic mtDNA was shown to be different from that of nDNA and to occupy only a small part of the total cellular DNA.

A study of higher eukaryotic mtDNA was first reported by Rabinowitz et al. (1965) and Swift (1965). According to Rabinowitz, the buoyant density of mtDNA from chick embryo heart and liver is $1.707 \text{ g/cm}^3$ and that of nDNA is $1.698 \text{ g/cm}^3$. Swift found a DNA component in sugar beet root mitochondria with a buoyant density of $1.705 \text{ g/cm}^3$ (the nDNA density was $1.689 \text{ g/cm}^3$). At the same time, several authors isolated cpDNA from tobacco leaves and reported it to be heavier in density than the nDNA (Shipp et al. 1965; Green and Gordon 1966; Tewari and Wildman 1966).

Thus, on the grounds of studies carried out from 1963 to 1966, it could be concluded that the minor DNA fractions found in the total preparations of eukaryotic DNAs are localized in cytoplasmic organelles – plastids and mitochondria.

A measure of theoretical incompliance and doubt was caused by a dAT component detected in crustaceans. It was difficult to conceive that this simple in structure deoxypolynucleotide is a mtDNA and that it performs any function in mitochondrial biogenesis.

Several reports published in 1966–1967 necessitated a cardinal reevaluation of the existing views concerning minor DNA distribution in cell organelle fractions. Borst and Ruttenberg (1966) as well as Corneo et al. (1966) found that in a number of cases nDNA itself is heterogenous and contains a stDNA and, consequently, is characterized by a bimodal distribution in the CsCl gradient. In a number of cases the mtDNA density coincides with that of the nDNA or coincides with the major component of nDNA if it contains a stDNA. According to Beridze et al. (1967), 30 % of the DNA isolated from bean leaf cell nuclei, purified in the sucrose gradient, was found in the zone of the density satellite component, $\varrho = 1.703 \text{ g/cm}^3$ (Fig. 4). The thermal denaturation curve of bean DNA showed a biphase shape confirming the data of density gradient ultracentrifugation on the presence of a minor DNA.

Subsequently, organelle DNA fractions with a density both coinciding with and differing from the nDNA density were excluded from the stDNA class and termed "organelle DNAs". DNA fractions of nuclear origin are designated as satellite DNAs.

About the same time Birnstiel et al. (1968) reported that rRNA genes from *Xenopus laevis* also form a discrete band in the CsCl gradient ($\varrho = 1.723 - 1.724 \text{ g/cm}^3$) considerably differing in density from the bulk of nDNA ($\varrho = 1.698 \text{ g/cm}^3$). A similar situation was observed later in other organisms (Patterson and Stafford 1971; Gall 1974; Blin et al. 1976; Cramer et al. 1976). The rRNA genes forming a separate band in the CsCl gradient are usually termed rDNAs.

Thus, the boundaries of the term satellite DNA have once more narrowed down to indicate minor fractions of nDNA with unknown functions.

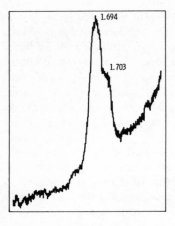

**Fig. 4**. Distribution of kidney bean leaf nDNA (*Phaseolus vulgaris*) in the CsCl density gradient (Beridze et al. 1967)

It should be noted that stDNA studies have become more intensive after the development of new methods of preparative equilibrium ultracentrifugation in the density gradient. Flamm et al. (1966) compared the distribution of DNA molecules in the CsCl gradient in angle and bucket rotors. Angle rotors give a five- to ten-fold better resolution than bucket rotors. The advantage of the angle rotor becomes self-evident since each tube can hold a ten-fold more amount of DNA than the bucket rotor tube. Namely, the use of angle rotors for preparative gradient ultracentrifugation has contributed to achievements in the field of stDNA research.

In the last years malachite green affinity resin has been used for isolation of stDNA and satellite chromatin (Bünemann and Müller 1978; Weber and Cole 1982a). The resin represents cross-linked bisacrylamide to which AT-specific dye, malachite green, is covalently attached by spacer of polyacrylamide chain. The resolution of this resin is sufficient for separation of two DNA species which differ by about 10% in their average GC content (Bünemann and Müller 1978).

## 2.2 Kinetic Satellites

The term satellite DNA was given quite a new meaning by Britten and co-workers (Waring and Britten 1966; Britten and Kohne 1966, 1968). Their work was very important for elucidating the general structural organization of an eukaryotic genome and revealed essential features of stDNA structure. Their most valuable result was the finding of the inverse proportional dependence between the size of a genome and the DNA reassociation rate for a wide range of organisms – from small viruses to animals. In these studies, measurements were made of the reassociation kinetics of DNA molecules under optimal conditions, such as the corresponding ionic strength, temperature, incubation time and size of DNA molecules. The controlling parameter of the reassociation reaction of separate DNA chains was the product of the DNA concentration and the incubation time. This parameter is designated by $C_0 t$ (Britten and Kohne 1968). The DNA of every organism or its separate fractions can be

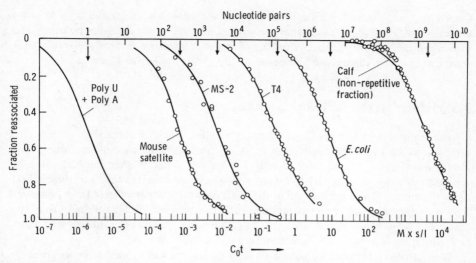

**Fig. 5.** Dependence of genome size on the $C_0t$ value for different double-stranded nucleic acid molecules (Britten and Kohne 1968)

characterized by the $C_0t$ value at which 50% of the reassociation reaction is completed under controlled conditions, $C_0t_{1/2}$ (Fig. 5).

Prior to Britten's work it was thought that the reassociation of higher organism DNAs was difficult to observe due to a high dilution of each individual nucleotide sequence. It was assumed that a reassociation reaction took months to complete. However, the very first investigations carried out by Britten and co-workers showed that some part of the DNAs of higher organisms reassociated very rapidly due to their high repetition in the genome. In mouse, the share of this fraction was 10%. It proved to be a stDNA detected earlier by Kit (1961) in mouse total DNA.

Mouse stDNA requires a higher $C_0t$ value for reassociation than a (poly A + poly U) double helix, wherein the number of nucleotide pairs is 1. However, mouse stDNA reassociation requires a lower $C_0t$ value than that for the replicative form of the MS-2 virus RNA consisting of $4 \cdot 10^3$ bp, for the T4 phage DNA ($2 \cdot 10^5$ bp) or, moreover, for the *E.coli* DNA. Calculations have shown that mouse stDNA consists of 300 bp long segments and is repeated about a $10^6$ times in every cell. If it is assumed that the length of a region with a satellite sequence is 30,000 bp ($20 \cdot 10^6$ daltons), then the satellite DNA will consist of 100 tandemly arranged repeats (Waring and Britten 1966). In the following years the method of reassociation kinetics demonstrated that other stDNAs consist of relatively short highly repetitive sequences. No highly repetitive sequences were observed in prokaryotic organisms; they are characteristic only for the eukaryotic genome.

The fact that stDNAs contain covalently linked and highly repetitive short sequences added is a new meaning to the term satellite DNA which is reflected in the definition of satellite DNA suggested by Walker (1971). According to Walker, a satellite DNA is a "native fraction of the chromosomal DNA which after isolation by any method gives a narrow unimodal band in CsCl because of common properties shared by its sequences". This definition does not include low molecular fractions into the satellite DNA class, since they do not exhibit narrow bands in the analytical ultracenrifuge. It also excludes duplexes formed at reassociation of separate DNA chains at low $C_0t$ values and belonging to the highly repetitive DNA class.

## 2.3 Hidden stDNAs; Twin Satellites

Up to the present we have dealt with density-satellite DNAs, but there exist "hidden" satellites which are not detectable in the CsCl neutral density gradient though they also consist of short repetitive sequences. This was what Walker implied when he spoke of isolating satellite DNAs by any method, without restricting it to the method of equilibrium ultracentrifugation in the CsCl density gradient. These fractions coincide in density with the major component and can be separated from the bulk of the major DNA by heavy metal (mercury, silver) ions or some antibiotics as well as by DNA-binding low molecular substances.

Ions of heavy metals, such as silver (Fig. 6) and mercury, are frequently used to separate hidden satellites from the bulk of DNA. The binding of heavy metal ions depends on the pH values. It has been shown that $Ag^+$ binding depends only on the base composition at pH $\leq 7$, while at pH $\geq 8$ it depends on the DNA base sequences (Macaya et al. 1978).

$Ag^+$ reversibly binds with DNA and can form two types of complexes. One is observable at $r_f < 0.2$ and depends on the presence of GC pairs. In this case silver ions are "introduced" between two base pairs, one of which is a GC pair. The other complex is formed in the range $r_f = 0.2 - 0.5$. Silver ions replace protons between thymine $N_3$ and adenine $N_1$ or between cytosine $N_3$ and guanine $N_1$. An increase of pH enhances the formation of this complex (Sissoëff et al. 1976).

Mercury ions also form two types of complexes. At low concentrations ($r_f < 0.5$) mercury ions replace two proteins in complementary bases and form N-Hg-N bridges. The AT pairs bind the mercury ions more intensively than the GC pairs and the denatured DNAs are more intensive than the native ones. With a further increase of concentration ($r_f > 0.5$) two mercury ions bind one base pair. The reversibility of the complex-forming reaction decreases with the increase of the pH value. The most effective condition for DNA molecule separation is $r_f \leq 0.5$ and pH 9 (Nandi et al. 1965).

The buoyant density of DNA increases considerably at $Ag^+$ or $Hg^{2+}$ binding. Chloride ions inhibit the binding of heavy metal ions, so $Cs_2SO_4$ is used as the gradient-forming substance instead of CsCl.

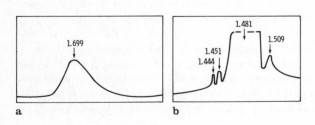

**Fig. 6.** Distribution of human DNA in the CsCl (**a**) and $Ag^+ - Cs_2SO_4$ (**b**) density gradients, demonstrating the presence of stDNAs: stDNA I (1.444), stDNA III (1.451), stDNA II (1.509) (Hearst et al. 1974; Corneo et al. 1971, (**a**) and (**b**), respectively)

Selective binding is used to create a large buoyant density difference between single- and double-stranded DNAs or between DNAs with different base compositions. For two λ phage chains the difference in density at $r_f = 0.3$ was $0.057\,g/cm^3$ in $Hg^{2+} - Cs_2SO_4$ and $0.009\,g/cm^3$ in CsCl. Thus, the separation of these two DNA chains in $Hg^{2+} - Cs_2SO_4$ as compared with CsCl is sixfold greater in density and four fold in distance under the same centrifugal forces (Nandi et al. 1965).

This method is utilized to increase the distances separating the different DNA fractions in the density gradient, as it essentially facilitates the procedure of minor DNA fraction isolation.

The use of heavy metal ions to isolate "hidden" stDNAs has been explained by the following reasons: (1) stDNAs bind more ions than required by the base composition; or, (2) stDNAs bind the amount of heavy metal ions corresponding to the base composition, but they manifest a greater buoyant density increment than DNAs with unique sequences (Sissoëff et al. 1976).

In addition to the above explanation it should be noted that in a number of cases stDNAs differ from the main DNA in base composition, but coincide with it due to the presence of $m^5C$ which decreases the stDNA buoyant density and it can by chance coincide with that of the main DNA.

StDNAs are considerably variable in density: they can be heavier or lighter than the main DNA or coincide with it in density. In some cases different satellites can have the same density, for example, stDNA from HS-α and HS-β of the kangaroo rat *Dipodomys ordii* (Hatch and Mazrimas 1974).

In certain cases a simultaneous combination of both $Ag^+$ and $Hg^{2+}$ ions can produce better results in the separation of individual DNA fractions. Using this method, Skinner and Beattie (1973) separated the stDNA of the hermit crab *Pagurus pollicaris* into its constituent fractions, and called them "twin satellites". Used separately, $Hg^{2+}$ and $Ag^+$ failed to yield the desired result.

Later, hidden stDNA was obtained not only with heavy metal ions in the $Cs_2SO_4$ gradient, but also with several antibiotics in the CsCl density gradient (Peacock et al. 1974; Manuelidis 1977). Netropsin, an antibiotic of an oligopeptide nature, has an affinity for AT pairs, while actinomycin D predominantly binds with the GpC sequences. Equilibrium ultracentrifugation of DNA preparations in their presence increases the resolution capability of the gradient and facilitates the isolation of stDNA as discrete fractions. Binding with antibiotics induces a decrease of the buoyant density of DNA molecules. Several organic substances binding with DNA and, as a rule, decreasing density in the neutral CsCl density gradient are used to an analogous end. Good results were obtained in isolating mouse stDNA with Hoechst 33258 acridine dye which specifically binds with AT-rich DNA sequences and increases resolution in the CsCl density gradient. BAMD−3,6-bis(acetatomercurimethyl)dioxane was successfully used to separate calf twin satellites (Kopecka et al. 1978).

Total plant satellite DNA representing a set of three types of molecules, stDNA, rDNA, and mtDNA, yielding one peak in the CsCl density gradient was separated into constituent components using the actinomycin D-CsCl density gradient (Fig. 7; Beridze 1980b).

The definition of satellite DNAs suggested by Britten et al. (1974) seems to precisely reflect the state of the art at that time: "Satellites: originally a minor component separated from the principal DNA in CsCl equilibrium density gradient centrifu-

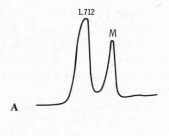

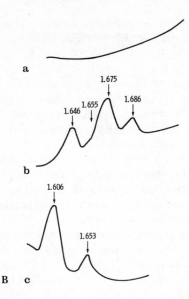

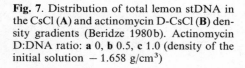

**Fig. 7.** Distribution of total lemon stDNA in the CsCl (**A**) and actinomycin D-CsCl (**B**) density gradients (Beridze 1980b). Actinomycin D:DNA ratio: **a** 0, **b** 0.5, **c** 1.0 (density of the initial solution $-$ 1.658 g/cm$^3$)

gation. Now also resolved with silver or mercury in Cs$_2$SO$_4$ gradients, or possibly merely exhibiting some of the properties which have come to be associated with the satellite class. These properties can be rapid and relatively precise reassociation, very many copies, simple sequences, homogeneous composition, purine/pyrimidine asymmetry between strands, concentration in centromeric heterochromatin, under-replication in polytene DNA, or occurrence in tandemly repeated clusters".

## 2.4 Conditions Requisite for stDNA Detection by Gradient Technique

According to modern concepts, the DNA of every eukaryotic chromosome is a single continuous chain (Kavenoff et al. 1973; Lauer and Klotz 1975). Consequently, it would be impossible to discover any stDNA if such long molecules were to be operated with. Usually, DNA molecules are fragmented without any special precautions during isolation procedure due to hydrodynamic shearing and the molecular mass of the DNA preparations amounts to $20-30 \cdot 10^6$ daltons.

The quantity of stDNA observed in total preparations depends on the length of the satellite region in the native molecule, and also on the size of the extracted DNA molecules. Fragmentation of a continuous DNA strand into relatively small pieces, yields fragments containing both pure satellite sequences and those consisting simultaneously of parts of the main and satellite components (Fig. 8).

This problem was discussed in detail by Brutlag and coauthors (Brutlag et al. 1977a). It is assumed that the satellite region m $-$ bp in length is surrounded by sequences with an essentially different density. The following conditions are required to obtain stDNAs in the discrete form:

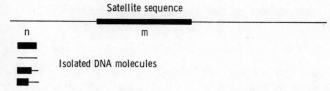

**Fig. 8.** The satellite region of the chromosomal DNA. Number of base pairs: m in the satellite region, n in the isolated DNA

(1) DNA fragmentation down to a defined size (n − bp in length) less than that of the satellite region length (n < m).
(2) Fragments flanking the satellite region must not appear in the density zone of the satellite component.
(3) Fragmentation must proceed uniformly so that equal amounts of molecules form at the end of the satellite region:

without flanking DNA, + 1 base pair of the flanking DNA, + 2 base pairs, + 3 base pairs, ... + (n − 1) base pairs, and, correspondingly, with a decreasing amount of satellite sequence. The total number of base pairs in "recombinant" fragments is

$$\frac{1}{n} \sum_{i=0}^{n-1} i = \frac{n-1}{2} \approx \frac{n}{2}$$

with $n \gg 1$. Since each satellite region has two ends, n base pairs of the satellite region will be lost. The portion of the fraction with the satellite density (f) will then be $f = (m − n)/m$.

Thus, the dimension of region m is proportional to the size of fragment n and inversely proportional $(1 − f)$ if $n < m$; $(1 − f)$ is the portion of recombinant fragments.

The second problem consists in the proportional extraction of the satellite and major DNAs. Schildkraut and Maio (1968) demonstrated that during extraction of mouse cell nuclei with 2 M NaCl the stDNA is completely retained in the insoluble material though the DNA yield is 80%.

According to Skinner and Triplett (1967) the dAT component of crustaceans is lost when DNA is isolated with hot phenol as the deproteinizing agent. Rae et al. (1976) observed that the ionic strength of the homogenization medium plays an important role in the isolation of *Drosophila* DNA with chloroform as a deproteinizing agent. If homogenization is performed in a NaCl solution with a concentration lower than 0.12 M the stDNA yield decreases.

These experiments indicate that chromatin is selectively sensitive to the ionic medium. The obtained data can be explained either by the specificity of stDNA binding with chromosome proteins or by its bonds with defined proteins which are not characteristic of the remaining part of the chromatin.

## 2.5 Contemporary Interpretation of the Term "Satellite DNA"

The following definitions of stDNA are encountered in the recent literature:

(1) Density (or patent) satellites yield a separate band in the CsCl density gradient.
(2) Hidden (or cryptic) satellites are detectable in the $Hg^{2+}$, $Ag^+ - Cs_2SO_4$ density gradient and also with the use of some antibiotics and DNA-binding low molecular weight substances in the CsCl density gradient. In the pure CsCl density gradient their density coincides with that of the major component.
(3) Twin satellites are two or more satellite DNAs coinciding in density in the neutral CsCl density gradient.
(4) Determination of the primary structure of the stDNA repetitive element in a number of organisms has given rise to a new classification of this class of molecules. StDNAs with a "simple" sequence consist of short nucleotide segments, while stDNAs with typically long repetitions are defined as "complex" ones (Rosenberg et al. 1978). A typical example of complex satellite DNAs is the α-stDNA of the African green monkey, where the length of the basic repeat unit is 172 bp and the 1.688 satellite from *Drosophila melanogaster* where the length reaches 359 bp (Rosenberg et al. 1978; Brutlag et al. 1978).

It is noteworthy that *D.melanogaster* contains both simple stDNAs (1.672, 1.686, 1.705) as well as a complex one (1.688).
(5) During the last years the term "satellites" has been used to describe tandemly arranged highly repetitive sequences regardless of whether or not they are separable from bulk DNAs by gradient ultracentrifugation (Pech et al. 1979a; Lee and Singer 1982; Witney and Furano 1983; Meyerhof et al. 1983). At limited digest of total nDNA by any restriction endonuclease these sequences generate a characteristic ladder of fragments that are demonstrable upon gel electrophoresis and are integral multiples of the basic repeat unit (Fig. 9). It is sometimes possible to select such a restriction endonuclease which digests nDNA leaving the tandemly arranged repeats in the high molecular form as demonstrated in the case of rat satellite I (Pech et al. 1979a). The satellites could be called restriction satellites, in contrast to stDNA demonstrable by gradient technique (gradient satellites).

It must be noted that the definition of restriction satellites needs clarification. According to what principle must one or another sequence be assigned to the satellite class – by its molecular mass or by the number of repeat units? For example, the 1400 bp sequence can quite rightly be called a satellite if the basic repeat is 2–10 bp in length, while in dealing with *Xenopus laevis* stDNA with a repeat unit length of about 700 bp, it will be just a dimer. We will be justified to speak of such a sequence as a satellite if at least ten or more such units are tandemly repeated.
(6) In recent years the terms "satellite" and "satellite-like" have been used to also denote short tandemly arranged simple sequences whose presence is observed at sequencing of cloned genes and their flanking regions (Emery and Weiner 1981; Maroteaux et al. 1983). These sequences can be arranged both near the 5'- or 3'-termini of a defined gene or within introns. These sequences may be without any restriction sites and so they are not detected by restriction analysis. They are not observed by the gradient method due to their low molecular mass. It seems

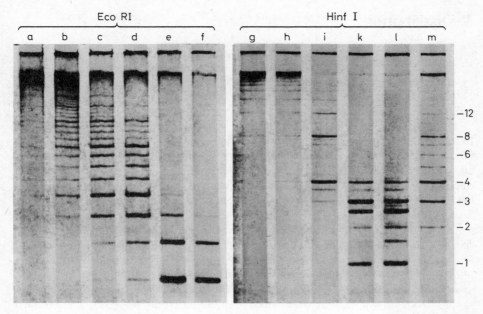

**Fig. 9.** Course of rat stDNA I digestion with increasing amounts of *Eco*RI (a–f) and *Hind*III (g–m). Electrophoresis in 1.5% agarose gel. The monomer and integral multiples of the 92/93 bp unit are indicated by *numbers* (Pech et al. 1979 a)

necessary to give them some new designation reflecting their states in the genome. Relatively short repetitive sequences 100–200 bp in length cannot be equated with true stDNAs dozens of kilobase pairs in length localized in centromeric heterochromatin.

In all, a more precise classification should be considered not only of stDNAs, but also of other fractions of the genome material. The creation of a distinct nomenclature similar to that of enzymes must be foreseen.

# 3 Distribution

In discussing problems concerning stDNA distribution, two questions must be answered. The first can be formulated as follows: do all eukaryotic species, both lower and higher, contain stDNAs? It is difficult as yet to give a final answer to this question. Published data permit to draw a conclusion that all eukaryotes contain hrDNAs. However, usually only part of these sequences is organized into tandem repeats – stDNAs.

Up to the present there has been no report in the literature on a study devoted specially to the detection of any density, hidden, or restriction forms of stDNA, and aimed to give negative results. Most likely the opposite can be assumed. With the use of modern experimental approaches apparently some form of stDNA can be detected in any eukaryotic species. Evidence of this is the detection by restriction endonucleases of stDNAs in lower eukaryotes – in two representatives of the genus *Trypanosoma* (Sloof et al. 1983).

The second question concerns the amount of stDNAs contained in one or another species. There is no precise data on this. The matter is that the content of all the density, hidden, and restriction stDNA forms must be determined in order to evaluate the percentage content of stDNA in any species. Mainly density stDNAs were analyzed in earlier studies where interspecies distinction was observed. As the analyzed species can contain other stDNA forms also, it is evident that the presented results in a number of cases could be underestimated. In the last years interest in quantitative determinations has significantly declined due to the use of cloning and sequencing techniques for analysis of stDNAs.

As mentioned a sufficiently large number of higher eukaryotic species have been studied in early papers with regard to their density stDNA content. Arrighi et al. (1970) analyzed the distribution of total DNA in the CsCl density gradient of 93 animal species, of which 48 species exhibited density stDNAs.

Table 1 presents the buoyant density values of the stDNAs detected in these animals. These authors observed cases when one species of a genus contained stDNA, while another of the same did not. These examples are also included in Table 1.

In calculating the buoyant density of DNA molecules, it is important to indicate the buoyant density of a reference DNA used for calculating the analyzed DNA density. The point is that in the first density-gradient ultracentrifugation studies the buoyant density of *E.coli* DNA was taken to be $1.710 \text{ g/cm}^3$ (Schildkraut et al. 1962).

Although the density of *E.coli* DNA was shown later to be $1.7035 \text{ g/cm}^3$ (Szybalsky 1968), the earlier value is still used in many studies with the respectively calculated densities of reference DNAs. This should be taken into account in calculations and in comparing data of different authors. A corresponding correction must be made in the empirical formula for the DNA GC-composition calculation based on the buoyant density value.

**Table 1.** Density stDNA content in animals (Arrighi et al. 1970)

| Object | Density, g/cm³ | | | Tissue source[a] |
|---|---|---|---|---|
| | Heavy stDNA | Major band | Light stDNA | |
| Insectivora | | | | |
| *Soricidae* | | | | |
| *Notiosorex crawfordi* | | | | |
| (Desert shrew) | | 1.699 | 1.684 | B |
| Chiroptera | | | | |
| *Molossidae* | | | | |
| *Molossops greenhalli* | | 1.6995 | 1.686 | B |
| *Phyllostomidae* | | | | |
| *Artibeus lituratus* | | 1.6985 | 1.684 | B |
| *Carollia perspicillata* | 1.7078 | 1.6976 | 1.6835 | B, C |
| *Chiroderma villosum* | | 1.6985 | 1.6835 | B |
| *Choeroniscus* (intermedius?) | | 1.6983 | 1.684 | B |
| *Vespertilionidae* | | | | |
| *Myotis keeni* | | 1.7005 | | C |
| *M.lucifugus* | | 1.699 | | B |
| *M.velifer* | 1.705 | 1.6990 | | C |
| Primates | | | | |
| *Cebidae* | | | | |
| *Alouatta caraya* | 1.701 | 1.698 | | A |
| (Howler monkey) | | | | |
| *A.villosa* | | 1.6975 | | A |
| (Howler monkey) | | | | |
| *Tupaiidae* | | | | |
| *Tupaia chinensis* | 1.704 | 1.6991 | | C |
| *T.glis* | 1.703 | 1.6985 | | C |
| *T.longipes* | | 1.6985 | | C |
| *T.minor* | 1.705 | 1.6985 | 1.683 | C |
| *T.montana* | 1.702 | 1.6986 | | B, C |
| *T.palawanensis* | | 1.6985 | | C |
| *Urogale everetti* | 1.705 | 1.6993 | 1.685 | C |
| Rodentia | | | | |
| *Caviidae* | | | | |
| *Cavia* sp. | 1.704 | 1.6971 | 1.682 | B, C, E |
| (Guinea pig) | | | | |
| *Cricetidae* | | | | |
| *Arborimus longicaudus* | 1.715 | 1.700 | 1.688 | C |
| (Red tree mouse) | | | | |
| *Clethrionomys rufocanus* | | 1.6995 | | B |
| (Red-backed mouse) | | | | |
| *C.mukuge* | | 1.6995 | 1.683 | B |
| (Red-backed mouse) | | | | |
| *Neotoma albigula* | | 1.701 | 1.6825 | C |
| (Wood rat) | | | | |
| *Muridae* | | | | |
| *Mus musculus* | | 1.6998 | 1.6898 | C |
| (House mouse) | | | | (continued) |

[a] A – cells from skin culture; B – cells from lung culture; C – liver; D – spleen; E – pancreas; F – cells from muscle culture; G – thymus

**Table 1** (continued)

| Object | Density, g/cm$^3$ | | | Tissue source[a] |
| --- | --- | --- | --- | --- |
| | Heavy stDNA | Major band | Light stDNA | |
| *Sciurudae* | | | | |
|   *Citellus tridecemlineatus* | | 1.698 | 1.683 | B |
|   (Ground squirrel) | | | | |
|   *Glaucomys volans* | | 1.6978 | 1.681 | C |
|   (Flying squirrel) | | | | |
|   *Tamiasciurus hudsonicus* | 1.7048 | 1.6965 | | B, C |
|   (Red squirrel) | | | | |
| Carnivora | | | | |
| *Canidae* | | | | |
|   *Canis familiaris* | | 1.6975 | 1.686 | C |
|   (Dog) | | | | |
| Felidae | | | | |
|   *Felis caracal* | | 1.697 | 1.682 | A |
|   (Caracal) | | | | |
|   *F.catus* | | 1.6973 | | B, C |
|   (House cat) | | | | |
|   *F.nigripes* | | 1.697 | | A |
|   (Black footed cat) | | | | |
|   *F.pardalis* | | 1.6973 | 1.684 | A |
|   (Ocelot) | | | | |
| Mustelidae | | | | |
|   *Amblonyx cinerea* | | 1.6975 | 1.682 | A |
|   (Otter) | | | | |
|   *Mephitis mephitis* | 1.7012 | 1.6978 | 1.683 | A, C, F |
|   (striped skunk) | | | | |
|   *Mustela erminea* | 1.7035 | 1.698 | | C, D |
|   (Ermine) | | | | |
|   *M.frenata* | 1.7045 | 1.6995 | | C, D |
|   (Long-tailed weasel) | | | | |
|   *M.putorius furo* | 1.707 | 1.6973 | | B |
|   (Ferret) | | | | |
|   *Spilogale putorius* | 1.7092 | 1.6976 | 1.682 | B, F, G |
|   (Spotted skunk) | | | | |
| Procyonidae | | | | |
|   *Bassariscus astutus* | | 1.697 | 1.683 | A |
|   (Ringtail) | | | | |
|   *Potus flavus* | | 1.697 | 1.685 | A |
|   (Kinkajou) | | | | |
|   *Procyon lotor* | | 1.6973 | 1.683 | A |
|   (Raccoon) | | | | |
| Ursidae | | | | |
|   *Ursus americanus* | 1.711 | 1.697 | | A |
|   (Black bear) | | | | |
|   *U.arctos* | 1.712 | 1.699 | 1.6835 | A |
|   (Syrian bear) | | | | |
| Viverridae | | | | |
|   Arctictis binturong | | 1.6973 | 1.6835 | A |
|   (Binturong) | | | | (continued) |

[a] A – cells from skin culture; B – cells from lung culture; C – liver; D – spleen; E – pancreas; F – cells from muscle culture; G – thymus

**Table 1** (continued)

| Object | Density, g/cm$^3$ | | | Tissue source[a] |
| --- | --- | --- | --- | --- |
| | Heavy stDNA | Major band | Light stDNA | |
| *Atilax paludinosus* (Marsh mongoose) | | 1.6975 | 1.6835 | A |
| *Bdeogale* sp. (Black-legged mongoose) | | 1.697 | 1.6835 | A |
| *Genetta* sp. (Genet) | | 1.6978 | 1.6833 | A |
| *Ichneumia albicauda* (White-tailed mongoose) | 1.712 | 1.698 | | A |
| *Suricata suricatta* (Meerkats) | | 1.6975 | 1.683 | A |
| *Viverricula indica* (Oriental civet) | | 1.6975 | 1.684 | A |
| Perissodactyla | | | | |
| *Equidae* | | | | |
| *Equus burchelli* (Zebra) | | 1.6965 | 1.682 | A |
| *E.caballus* (Horse) | 1.7107 | 1.6971 | 1.682 | A, C |
| *E.przewalskii* (Mongolian wild horse) | 1.711 | 1.6973 | 1.686 | A |
| *E.zebra hartmannae* (Zebra) | 1.705 | 1.696 | | A |
| Artiodactyla | | | | |
| *Bovidae* | | | | |
| *Ovibos moschatus* (Musk ox) | 1.712 | 1.6985 | 1.6825 | A |
| *Cervidae* | | | | |
| *Dama dama* (Fallow deer) | 1.708 | 1.697 | | A |
| *Odocoileus virginianus* (Whitetail) | 1.711 | 1.6985 | | A |

[a] A – cells from skin culture; B – cells from lung culture; C – liver; D – spleen; E – pancreas; F – cells from muscle culture; G – thymus

According to Arrighi et al., light stDNAs with the density 1.681–1.686 g/cm$^3$ result from microbial contamination of the objects. Mycoplasms are cultivated in many of the cell cultures used in the studies. The DNA isolated from pure mycoplasm cultures occupies the given region in the density gradient.

On the other hand, it is not excluded that some stDNAs detected by Arrighi et al. are indeed mtDNAs, since it was total DNA that was analyzed. It is known that the buoyant density of animal mtDNA varies in a wide range of 1.681–1.711 g/cm$^3$ (Gauze 1977).

An important parameter characterizing stDNAs is their percentage content in the genome, but unfortunately such data are not provided by the author.

Ingle et al. (1973) studied DNA distribution in 71 higher plant species in the CsCl density gradient. Of the 58 analyzed species of dicotyledonous plants, 27 contained a density satellite component (Table 2). Of the 27, only two stDNAs, that of *Lobularia*

**Table 2.** Density stDNA content in dicotyledonous plants (Ingle et al. 1973)

| Order No.[a] | Family | Species | Common name | Buoyant density | | |
|---|---|---|---|---|---|---|
| | | | | Major component | stDNA | stDNA (%) |
| 1 | Winteraceae | Drimys piperita | | 1.698 | 1.709 | 7 |
| 16 | Hamamelidaceae | Hamamelis mollis | Witch hazel | 1.695 | 1.706 | 6 |
| 36 | Cucurbitaceae | Cucumis melo | Melon | 1.692 | 1.706 | 25 |
| | | Cucumis sativus | Cucumber | 1.694 | 1.702 | 28 |
| | | | | | 1.706 | 16 |
| | | Cucurbita pepo | Marrow | 1.696 | 1.706 | 18 |
| | | | Pumpkin | 1.695 | 1.707 | 16 |
| | | | Squash | 1.695 | 1.706 | 17 |
| | | Citrullus vulgaris | Watermelon | 1.693 | 1.708 | 3 |
| | | Bryonia dioica | White bryony | 1.696 | 1.707 | 5 |
| | | Lagenaria vulgaris | Bottle gourd | 1.692 | 1.707 | 9 |
| | | Luffa cylindrica | | 1.696 | 1.707 | 6 |
| 38 | Cruciferae | Brassica rapa | Turnip | 1.696 | 1.704 | 21 |
| | | Brassica pekinensis | Chinese cabbage | 1.695 | 1.703 | 17 |
| | | Lobularia maritima | | 1.695 | 1.688 | 23 |
| | | | | | 1.706 | 9 |
| 50 | Leguminosae | Phaseolus coccineus | Runner bean | 1.693 | 1.702 | 24 |
| | | P.vulgaris | French bean | 1.693 | 1.703 | 19 |
| | | P.aureus | Mung bean | 1.692 | 1.705 | 5 |
| 56 | Rutaceae | Citrus sinensis | Sweet orange | 1.694 | 1.712 | 23 |
| | | Citrus limonia | Lemon | 1.694 | 1.711 | 23 |
| | | Citrus paradisi | Grapefruit | 1.693 | 1.711 | 23 |
| | | Citrus nobilis | Tangerine | 1.693 | 1.712 | 19 |
| | | Fortunella sp. | Kumquat | 1.693 | 1.712 | 24 |
| 58 | Linaceae | Linum usitatissimum | Flax | 1.699 | 1.689 | 15 |
| | | Linum grandiflorum rubrum | Red flax | 1.698 | – | 0 |
| 70 | Solanaceae | Solanum tuberosum | Potato | 1.695 | 1.707 | 4 |
| | | Solanum crispum | | 1.698 | 1.710 | 6 |
| | | Solanum capsicastrum | Christmas orange | 1.693 | – | 0 |
| | | Lycopersicon esculentum | Tomato | 1.694 | 1.705 | 8 |
| 74 | Compositae | Calendula officinalis | Pot marigold | 1.692 | 1.705 | 11 |

[a] Order No. is taken from the classification of Takhtajan (1966)

maritima ($\varrho = 1.688$ g/cm$^3$) and *Linum usitatissimum* ($\varrho = 1.689$ g/cm$^3$) were lighter in density than the major DNA; the others were heavier ($\varrho = 1.702–1.712$ g/cm$^3$). The highest proportion, 28% of the genome, was observed in one of two stDNAs from *Cucumis sativus* ($\varrho = 1.702$ g/cm$^3$).

The authors noted the absence of stDNAs in all of the 13 analyzed monocotyledonous plants. However, stDNA was later detected in a monocotyledonous plant – the orchid *Cymbidium ceres* (Capesius et al. 1975).

It has been noted that density stDNAs are the most typical for two families, *Cucurbitaceae* and *Rutaceae*, although species without stDNAs have been observed in both of them (Ingle et al. 1973).

In the above discussed studies only density stDNAs were considered by the authors. Those species, the DNAs of which are characterized by a typical unimodal distribution, can contain hidden satellites. Moreover, a thorough analysis can separate some stDNAs into their constituent components. Calf DNA can be cited as an obvious example. As mentioned above, originally only asymmetry in the DNA peak "heavy shoulder" was observed in the CsCl gradient. At present, eight stDNAs have been distinguished in calf DNA using different density gradient ultracentrifugation methods (Kopecka et al. 1978).

In general, it should be noted that there is no interrelation between density stDNA content and the conventional schemes of animal and plant evolution or with their taxonomic classification.

Lower eukaryotic organisms have been insufficiently studied with regards to their stDNA content. A minor DNA fraction with a $1.704 \text{ g/cm}^3$ density (major component density $1.700 \text{ g/cm}^3$) was detected in yeast *Saccharomyces cerevisiae* as early as 1966 (Corneo et al. 1966). This DNA was neither isolated in the pure form nor characterized. As mentioned above, stDNAs were detected in two protozoan species of the genus *Trypanosoma*.

# 4 Reassociation Kinetics

Reassociation of separate DNA strands is a second-order reaction described by the formula

$$(C_0 - C)/C = k_2 C_0 t,$$

where $C_0$ and $C$ are the concentrations of the reacting chains in solution at zero time and time t, respectively. After half of the reassociation reaction has passed, i.e., $C_0/C = 2$, then $k_2 = 1/C_0 t_{1/2}$. Thus, the second-order reaction constant is inversely proportional to the $C_0 t_{1/2}$ value.

The process of reassociation, i.e., duplex formation, can be defined by two methods: (1) hydroxyapatite chromatography to separate double- and single-stranded DNA molecules; (2) spectrophotometrically, monitoring duplex formation by the decrease of UV-light absorption at 260 nm.

The rate of duplex DNA formation is also measured by hydroxyapatite chromatography, with the single-stranded DNA loops being previously removed by S1 nuclease.

If the first method (hydroxyapatite chromatography) is used to estimate the amount of formed duplexes, the shape of the kinetic curves in all the $C_0 t$ range corresponds to the second-order reaction. Hydroxyapatite separates single-stranded DNA molecules from the others, including imperfect duplexes. In this method each formed duplex is assumed as an act of the reaction.

However, the optical method, just as hydroxyapatite chromatography, after preliminary treatment of the reaction products with S1 nuclease, detects fractions representing true double helices, the loops in the imperfect duplexes being excluded from the reaction products.

It has been demonstrated that the rate of formation of duplex DNA measured by S1 nuclease considerably deviates from the second-order reaction (Britten and Davidson 1976). DNA reassociation measured spectrophotometrically is described as a second-order reaction approximately up to the moment when one-third of all the DNA has been reassociated (Gillis et al. 1970).

On the basis of the reassociation rate constant $k_2$ in kinetic experiments, one can calculate the kinetic complexity of DNA molecules, i.e., the number of the nucleotide pairs in non-repeating DNA sequences, using the formula

$$N = 3 \cdot 10^5 L^{1/2}/k_2$$

where N is the kinetic complexity, L is average number of base pairs in single-stranded DNA molecules (DNA fragment length). The reaction proceeded at $(T_m - 25)\,°C$ and $1.0\,M\,Na^+$ (Wetmur and Davidson 1968).

The suggested equation for kinetic complexity calculation gives an exact value only for unique DNA sequences. In analyzing the structural organization of guinea pig α-stDNA, Southern (1970) noticed that the chemically determined basic sequence (6 bp) is considerably less than the kinetic complexity derived from reassociation kinetics ($10^5$ bp). Similar discrepancies were observed for other stDNAs as well (Botchan 1974; Salser et al. 1976).

Southern (1971) suggested that one of the possible explanations of this disparity is the mismatching of separate chains due to satellite sequence divergency affecting the DNA reassociation rate. Southern proposed an equation to correct this effect

$$\log R_m/R_0 = n \log(1 - P)$$

where $R_m$ is the reassociation rate of the mismatched DNA duplexes, $R_0$ is the reassociation rate of the same DNA at formation of perfect duplexes (ideal reassociation), n is the number of base pairs required to stabilize nucleation in the conditions used for reassociation [the formation of several base pairs in a correct register (nucleation) is considered to be a rate-limiting step in the reassociation]. Subsequent base pairing after nucleation proceeds rapidly (zippering), P is the fraction of mismatched base pairs in the duplex.

Sutton and McCallum (1971) calculated n to be 13 bp for stDNA under reassociation conditions (0.12 M phosphate Na, 60°).

Using the above formula these authors corrected the $C_0 t_{1/2}$ values for various stDNAs (Table 3).

The proportion of mismatched bases was derived from the $T_m$ decrease of the reassociation hybrids as compared to that of the native DNA. Sutton and McCallum proceeded from the fact that a 20° decrease of $T_m$ of the reassociated mouse stDNA duplex induces a 20- to 50-fold decrease in the rate of reassociation. However, it was shown later in model experiments that the rate of reassociation is almost twice less at $\Delta T_m = 10°$ (Bonner et al. 1973; Hutton and Wetmur 1973). This is why probably Sutton and McCallum's calculations of the kinetic complexity of stDNA did not give the true length of repeats.

Hutton and Wetmur (1973) in a study of mouse stDNAs showed that the kinetic complexity of this satellite is a function of the lengths of the reacting chains and, consequently, that the kinetic complexity of stDNAs cannot be determined by the same method used for unique DNAs. Hutton and Wetmur obtained mouse stDNA fragments of different lengths. Spectrophotometric evaluation of the kinetic complexity of these fragments coincided with their length.

According to the authors, the process of nucleation at reassociation leads to a divergence from second-order kinetics and, with the increase in fragment length, to the corresponding increase of the calculated kinetic complexity.

**Table 3.** Corrected values of $C_0 t_{1/2}$ for some stDNAs (Sutton and McCallum 1971)

| DNA | $\Delta T_m$, °C | Observed $C_0 t_{1/2}$ | Corrected $C_0 t_{1/2}$ | Corrected kinetic complexity, base pairs |
|---|---|---|---|---|
| Guinea pig stDNA II | 3 | $4.1 \cdot 10^{-4}$ | $2.4 \cdot 10^{-4}$ | 130 |
| Mouse stDNA | 5 | $6.6 \cdot 10^{-4}$ | $2.6 \cdot 10^{-4}$ | 140 |
| Guinea pig stDNA III | 8 | $9.5 \cdot 10^{-4}$ | $2.1 \cdot 10^{-4}$ | 110 |
| Calf stDNA I | 10.5 | $5.0 \cdot 10^{-3}$ | $7.2 \cdot 10^{-4}$ | 390 |

In a qualitative study of these questions Chilton (1973) demonstrated that the conventional method of kinetic complexity calculation using the reassociation rate constant cannot be applied if the repeat length is less than that of the reacting DNA strands.

Using the available data Chilton showed that in the case considered, reassociation kinetics leads to a kinetic complexity value equal to the molecular weight of the analyzed fragment. The information obtained in such cases can only be evidence that the repeat length is shorter than that of the analyzed DNA fragment.

The empirical dependence of $k_2$ on the molecular weight for unique DNA is known to be

$$k_2 \approx M^{1/2}/N$$

where N is the kinetic complexity.

This relationship is derived as the result of two opposite effects of the $k_2$ dependence on the molecular weight. The first is that the amount of the duplex formed per successful nucleation is directly proportional to the molecular weight. The second is that the rate of duplex formation is inversely proportional to $M^{1/2}$ due to diffusional restrictions of the reaction.

In the case of a repeating DNA, $k_2$ does not manifest the usual dependence on the molecular weight. An increase of the molecular weight of reactant chains does not increase the length of a duplex formed per successful nucleation. The length of the formed duplex is proportional to the repeat length N. Thus, in the case of such sequences, the first factor (M) in this relationship is replaced by N with the resulting equation

$$k_2 \approx 1/N(NM^{-1/2}) \approx M^{-1/2}.$$

Consequently, $k_2$ does not depend on N.

Renaturation of such a repeating sequence will proceed faster and not slower with a decrease of the molecular weight. Namely, this was observed by Hutton and Wetmur (1973) for mouse stDNA.

A conventional analysis of DNA renaturation reaction fitting this model yields a kinetic complexity N, exactly corresponding to the fragment size M, but not to the length of the repeating unit.

Thus, studies of stDNA reassociation kinetics provide evidence that in the majority of cases the length of the repeat, calculated by the kinetic method, essentially exceeds the actual length of the repeat, and the corrected values in some instances differ significantly from the true length due to the absence of accurate criteria for determining the correction coefficients for stDNAs. If it is assumed that Chilton's estimates are true, this would mean that it is utterly impossible to estimate the repeat length in "simple" stDNAs by reassociation kinetics.

Although the reassociation kinetics method does not ensure an accurate estimation of the repeat lengths in simple stDNAs, it may help to detect the existence of repeats of different lengths in stDNAs.

Using the spectrophotometric method Marx and Hearst (1975) showed that mouse stDNA and the kangaroo rat HS-β-satellite consist of two components differing in kinetic properties and suggested a formula for analysis of two-component systems

$$\frac{d(1/A - A_\infty)}{dT}\bigg|_{t\to 0} = 2.04 \cdot 10^{-4}(k_f \theta_f^2 + k_s \theta_s^2)$$

where $\theta$ is the fraction of each componet, t is the time, f is the rapidly reassociating component, s is the slowly reassociating component, A is the absorbance at t, $A_\infty$ is the absorbance at infinite time.

The idea of the method is that at large t values the fast component no longer contributes to the $A_{260}$ decrease. A curve of the dependence of $1/(A - A_\infty)$ on the time is plotted. Extrapolation of the curve to $t = 0$ produces an inverse value of hyperchromicity of the slow component (Fig. 10).

As a result of the performed experiments, the authors estimated the kinetic complexity of separate components of a stDNA:

| stDNA component | $\theta$ | Kinetic complexity, bp |
|---|---|---|
| | Mouse | |
| f | 0.69 | $130 \pm 25$ |
| s | 0.31 | $1250 \pm 250$ |
| | Kangaroo rat | |
| f | 0.93 | $56 \pm 10$ |
| s | 0.07 | $240 \pm 50$ |

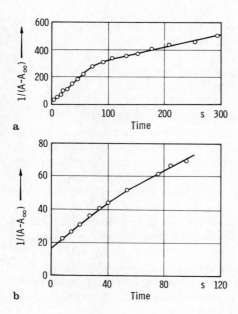

**a**

**b**

**Fig. 10.** Kinetics of optical reassociation of kangaroo rat stDNA (Marx and Hearst 1975). **a** 0.36 M Na$^+$; **b** 0.075 M Na$^+$

A primary structure analysis of the 234 bp repeat unit of mouse stDNA determined later by Hörz and Altenburger (1981) permitted to detect basic subrepeats of the stDNA 58–60 bp in length, consisting in turn of shorter 28–30 bp sequences. The 130 ± 25 bp unit detected by reassociation kinetics does not have a true analog in the 234 bp repeat. Half of the main 234 bp repeat observed at restriction analysis comprises a small part of the restricts. Consequently, the repeat lengths estimated by Marx and Hearst do not reflect the actual organization of the given stDNA.

# 5 Chromosomal Location

One of the decisive factors drawing together cytogenetical studies and those on the structural organization of chromosomal DNAs is the method of cytological hybridization permitting the detection of the localization sites on chromosomes of a number of repeating DNAs, including rRNA, tRNA, and histone genes as well as stDNA. The method was suggested by several research groups (Pardue and Gall 1969; John et al. 1969; Buongiorno-Nardelli and Amaldi 1970). The essence of the method is that DNA is denatured by any denaturating agent in metaphase chromosomes, applied on a glass slide. The chromosomes are then hybridized with radioactive nucleic acid (RNA or single-stranded DNA) under conditions similar to hybridization on membrane filters. The unhybridized molecules are removed with RNase or by washing with a buffer solution and the plates are covered with an autoradiographic emulsion. After development the preparations are stained and the localization sites of the silver grains on the chromosomes are determined, i.e., the sought for DNA sequences are identified.

The simplest experiments on cytological hybridization use total radioactive DNA or RNA transcribed in vitro from total DNA. Autoradiography with increasing exposure provides an approximate visualization of reassociation curves since the more repeating sequences are hybridized at lower exposure. Experiments with *Drosophila* salivary gland chromosomes showed that RNA, complementary to total DNA, is hybridized primarily with heterochromatin of the chromocenter, indicating that the most highly repeating sequences are concentrated in this region. Binding is subsequently observed with different parts of the chromosome arms reflecting the distribution of sites containing either a less amount of hrDNA or DNA sequences with a lower repeatability.

The first single DNA with an established localization was a mouse stDNA localized at centromeric heterochromatin in all chromosomes except the Y (Pardue and Gall 1969; Jones 1970) (Fig. 11). In investigations that followed this work, it was shown that localization of the majority of stDNAs is restricted by centromeric heterochromatin or the heterochromatic arms of chromosomes (Pardue 1975). Heterochromatin is usually divided into two classes. The regions of the constitutive heterochromatin are constantly condensed and stained more intensively than the remaining part of chromatin. Some chromosomes or their parts reveal heterochromatic properties only in certain tissues at defined stages of development and are regarded as facultative heterochromatin. Hybridization experiments in situ have shown that only the constitutive heterochromatin is hybridized with stDNAs.

Pardue and Gall (1972) reported a correlation of quantitative differences of centromeric heterochromatin and stDNA in related species. Moreover, the variability of the GC content of stDNAs in different species is a confirmation that different sequences can perform DNA functions in centromeric heterochromatin.

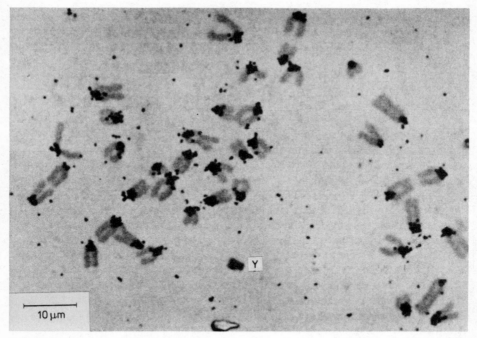

**Fig. 11.** Cytological hybridization of radioactive cRNA with mouse stDNA (Hearst et al. 1974)

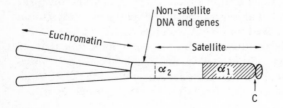

**Fig. 12.** Organization of *D.virilis* heterochromatin (Holmquist 1975). C–centromere

In centromeric heterochromatin itself the stDNAs are arranged in a specific manner. Thus, for example, *D. virilis* chromosomes contain stDNAs 1.671 and 1.688 in the heterochromatic region flanking the centromere ($\alpha_1$-heterochromatin) and in the short arm of chromosomes; $\alpha_2$-heterochromatin contains stDNA 1.692 in the proximal region and a non satellite DNA and genes in the distal region (Fig. 12).

It should be mentioned that hrDNAs are detected near centromeres also in organisms without density stDNAs. It has been shown elsewhere that in some cases stDNA localization is not limited by centromeric heterochromatin.

White et al. (1975) showed that mouse stDNA is detectable not only in centromeric heterochromatin, but in the long arm of some chromosomes as well. At staining those sites produced C-banding. Kurnit and Maio (1974) detected one of the three African green monkey satellites mainly outside the centromere. Such a list could be continued.

The C-banding of eukaryotic chromosomes should be briefly discussed in this chapter since the stDNA localization sites, as a rule, coincide with the C-bands in experiments in situ. The C-bands correspond to constitutive heterochromatin. The method was suggested by Arrighi and Hsu (1971) and included a consecutive treatment of chromosomes with HCl to remove part of the histones, with RNase to remove RNA molecules, with alkali to denature the DNA, DNA annealing and staining with Giemsa solution.

Apparently, denaturation of all the chromosomal DNA occurs at alkali treatment of chromosomes, while subsequent reassociation restores the original structure of only the hrDNA contained in heterochromatin. Therefore, Giemsa staining is more effective in the constitutive heterochromatin region.

# 6 Structural Features

## 6.1 Protozoa

### 6.1.1 *Trypanosoma* stDNA

StDNA has been detected in two representatives of the *Trypanosoma* genus, the African (*T.brucei*) and the South American (*T.cruzi*) (Sloof et al. 1983). StDNAs were isolated in the sucrose density gradient as rapidly sedimenting fractions after digestion of nDNA with certain restriction endonucleases.

The size of the *T.brucei* stDNA repeat unit is 177 bp and it consists of two copies of 19 bp sequences differing in 1 bp and several additional copies of part of this sequence (Fig. 13a). The length of the *T.cruzi* stDNA is 196 bp (Fig. 13b). The presence of short subrepeats is not detected by sequence analysis. No homology is exhibited between the *T.brucei* and *T.cruzi* stDNA sequences.

The detection of stDNA in the genus *Trypanosoma* is the first example of the presence of stDNA in lower eukaryotes. It is noteworthy that the genetic complexity of the indicated protozoans is less than ten times that of *Escherichia coli*.

## 6.2 Arthropods

### 6.2.1 Crustacean stDNA

Density stDNAs have been detected in most of the crustaceans (Skinner 1967; Skinner et al. 1970; Beattie and Skinner 1972) (Table 4) and can be divided provisionally into three groups. Together with the light dAT component, satellites of moderate density from 1.682 to 1.688 g/cm$^3$, as well as heavy 1.714 to 1.725 g/cm$^3$ satellites were identified.

#### 6.2.1.1 dAT Satellite

Sueoka (1961) demonstrated that a considerable part of the crab *Cancer borealis* and *Cancer irroratus* genome represents a light stDNA with a density of 1.677 g/cm$^3$ (Fig. 14). Waldfogel and Swartz (1971) isolated nuclei and mitochondria from *Cancer borealis* tissues and demonstrated the nuclear origin of the dAT polymer.

The in vitro replication of *C.borealis* stDNA, in contrast to synthetic dAT polymer, required the presence of all four deoxyribonucleoside-triphosphates (Swartz et al. 1962). Without dGTP and dCTP the rate of synthesis was only 19% as compared with a complete system. However, when dTTP was absent in the reaction mixture, the rate dropped to 0.1%. These facts evidence that some C and G residues

AluI
‾‾

<pre>
         10          20          30          40          50          60
CTAATAAATG  GTTCTTATAC  GAATGAATAT  TAAACAATGC  GCAGTTAACG  CTATTATACA
         -----C-      T-----     ---G-
           Sau961     TaqI       MboⅡ
           -A--
           MboI
</pre>

                                                        HhaI     HincⅡ

<pre>
         70          80          90         100         110         120
CAATAACTTT  TAATGTGTGC  AATATTAATT  ACAAGTGTGC  AACATTAAAT  ACAAGTGTGT
</pre>

                                                                  AluI
                                                                  ‾‾

<pre>
        130         140         150         160         170
AACATTAATT  TGCAAGTTTG  CAACGCTGTT  CTTTAGTGTT  TAATGTGTGC  AACAAAG
</pre>

**a**

AluI
‾‾

<pre>
         10          20          30          40          50          60
CTCGCGAAAT  TCCTCCAAGC  AGCGGATAGT  TCAGGGTTGT  TTGGTGTCCA  GTGTGTGAAC
                                   HinfI
         70          80          90         100         110         120
ACGCAAACAG  AYATTGACAG  AGAGTGCCTC  TGACTCCCRC  CATTCACAAT  CGCGAAACAA
                                                 MboI
        130     *   140         150         160         170         180
AAATTTGGAC  CACAACGTGT  GRTGCAGCGG  CCGCTCGAAA  ACGATCCGCC  GAGTGCAGCA
            AluI
            ‾‾
        190
CCCGTGTGGG  CAAGAG
</pre>

**b**

<pre>
2       10           20
TAA  TAAATGGTTCTTATACGAA  TGAATA
      ·  · ·· ·  ··     ·
       30        40
     TTAAACAATGCGCAGTTAA
       ·
       50        60
     CGCTATTATACACAATAAC  TT
        ·
     70        80
     TTAATGTGTGCAATATTAA  T
     ··· ·     ·   ·  ·


     TACAAGTGTGCAACATTAA  A

     110       120         130
     TACAAGTGTGTAACATTAA  TT
             ·
              140       150
     TGCAAGTTTGCAACGCTGT  TCTTT
        ·
     160       170      1
AGTGT  TTAATGTGTGCAACAAAGC
        ··· · · ·    ···· ·
</pre>

**c**

**Fig. 13.** Nucleotide sequence of *T.brucei* and *T.cruzi* stDNA. **a** 177 bp repeat of *T.brucei* stDNA; **b** 196 bp repeat of *T.cruzi* stDNA. Y and R denote pyrimidines (about 50% C/T) and purines (about 50% A/G), respectively. *Asterisk* denotes G 20%; **c** 19 bp repeat of *T.brucei* stDNA. Difference from the consensus sequence is denoted by *dots* (Sloof et al. 1983)

are interspersed into stDNA chain. Analysis of the synthesized product showed that the base ratio in the stDNA is $A:T:G:C = 0.84:1.07:0.030:0.028$. Consequently, the GC content is about 3%. *C.borealis* stDNAs were also studied by the nearest neighbor method (Table 5). Proceeding from these data it can be concluded that the *C.borealis* stDNA is structurally very close to that of the synthetic dAT copolymer, the consecutive alternation of the A and T residues being characteristic for 93% of

**Table 4.** Buoyant density of DNA components of crustaceans (data from Skinner et al. 1970; Beattie and Skinner 1972)

| Species | Major component | stDNA[a] | | |
|---|---|---|---|---|
| | | Light | Moderate density | Heavy |
| *Callinectes sapidus* | 1.696 | 1.677(15) | 1.688(< 1) | — |
| *Gecarcinus lateralis* | 1.700 | 1.677(18) | 1.688(< 1) | 1.721(3) |
| *Homarus americanus* | 1.699 | — | 1.688(< 1) | 1.715(10) |
| *Maja squinado* | 1.698 | — | 1.684(< 1) | — |
| *Pagurus pollicaris* | 1.701 | — | — | 1.725(< 1) |
| *Procambarus blandingii blandingii* | 1.699 | — | 1.682(< 1) | — |
| *Libinia dubia* | 1.698 | 1.675(6) | — | 1.714(2) |
| *Cancer borealis* | 1.698 | 1.676(23) | — | — |
| *Cancer pagurus* | 1.701 | 1.677(24) | — | 1.721(4) |

[a] Values in brackets show percentage of the stDNA

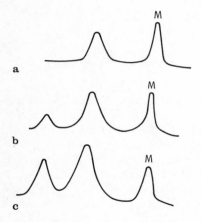

**Fig. 14.** Crab DNA distribution in the density gradient (Sueoka 1961). **a** *Carcinus maenas*; **b** *Cancer irrotatus*; **c** *Cancer borealis*, M $^{15}$N-DNA *Serratia marcescens*

**Table 5.** Nearest neighbor frequencies of *C.borealis* stDNA (Swartz et al. 1962)

| Sequence | Frequency of occurrence | Sequence | Frequency of occurrence |
|---|---|---|---|
| ApA | 0.0127 | GpT | 0.0081 |
| TpT | 0.0126 | ApC | 0.0069 |
| CpA | 0.0100 | GpG | 0.0009 |
| TpG | 0.0089 | CpC | 0.0009 |
| GpA | 0.0042 | TpA | 0.504 |
| TpC | 0.0015 | ApT | 0.429 |
| CpT | 0.0004 | CpG | 0.0007 |
| ApG | 0.0018 | GpC | 0.0015 |

the sequences. It should be also mentioned that all the 16 possible dinucleotides are observed in a different quantitative ratio.

Gray and Skinner (1974) applied the ultraviolet circular dichroism method to study the primary structure of three AT-rich stDNAs of crustaceans (*C.borealis*, *Gecarcinus lateralis*, *Callinectes sapidus*) with an identical density in the neutral CsCl density gradient. The ultraviolet circular dichroism spectra of a double-stranded DNA are determined both by their base composition and their distribution along the DNA chain. Many double-stranded DNAs with a simple base distribution, such as synthetic DNAs containing serially repeating dinucleotide and trinucleotide sequences, have a characteristic circular dichroism spectra.

In this study the circular dichroism spectra of three DNAs were compared with synthetic polynucleotide spectra to determine the frequency of nearest neighbor occurrence in DNA. It was shown that the *C.borealis* stDNA contains almost homogeneous $d(AT) \cdot d(AT)$ sequences; the *G.lateralis* stDNA contains less homogeneous poly $d(AT) \cdot d(AT)$ sequences and, probably, more $d(AA) \cdot d(TT)$, $d(AC) \cdot d(GT)$ and $d(TG) \cdot d(CA)$ dimers than the *C.borealis* stDNA.

The stDNA of *Callinectes sapidus* has very little or no $d(AT)_n \cdot d(AT)_n$. This DNA consists of tandemly repeating sequences of the dimer pairs $d(AA) \cdot d(TT)$, $d(AT) \cdot d(AT)$, and $d(TA) \cdot d(TA)$ and at the same time has G- and C-containing dimer pairs.

### 6.2.1.2 *P.pollicaris* stDNA

Beattie and Skinner (1972) have reported that the stDNAs from *H.americanus* and *P.pollicaris* ($\varrho = 1.715$ and $1.725 \text{ g/cm}^3$, respectively) show more than two bands in the alkaline CsCl gradient. This is evidence that these two fractions are twin satellites.

A heavy stDNA ($\varrho = 1.725 \text{ g/cm}^3$) constituting less than 1% of the total DNA was isolated from the hermit crab *Pagurus pollicaris* cells (Skinner and Beattie 1974). The stDNA was separated into two fractions (stDNA I and stDNA II) in the $Cs_2SO_4$ gradient in the simultaneous presence of $Ag^+$ and $Hg^{2+}$. One stDNA I chain contained 50% deoxyguanylate residues, being essentially free of deoxycytidylate ($< 3\%$). The other chain revealed 47% deoxycytidylate and less than 2% deoxyguanylate. In stDNA II the divergence in composition between the two strands is expressed to a less extent (dG : dC = 36 : 26% and 28 : 32%, respectively). The 1.725 twin satellites differ also in the total GC content (53% in stDNA I and 60% in stDNA II).

In a later study Skinner et al. (1974) determined the sequence of the hermit crab stDNA I repeating unit. To do so, cRNAs from separate DNA chains were obtained using RNA polymerase. The cRNAs were cleaved with RNase $T_1$ and pancreatic RNase. Owing to the structural simplicity, one or two oligonucleotides prevailed in every fingerprint. The repeating unit of *P.pollicaris* stDNA I has the following structure:

> 5'... TAGG
>      ATCC ... 5'.

Skinner et al. (1974) noted that this tetramer is included in the hexamer constituting 50% of the guinea pig α-stDNA sequence. There is also another similarity between stDNA I and the guinea pig α-stDNA in that one chain practically contains no cytosine and the other no guanine. Another similarity with *D.ordii* HS-α stDNA is also noted. The authors assume that this sequence can occur in other GC-rich animal satellites.

Later the same group determined the sequence of the repeating element in *P.pollicaris* stDNA II (Chambers et al. 1978):

$$5'\ldots\binom{CAG}{GTC}_n\ \frac{CTGCACT}{GACGTGA}\ldots 5'$$

where n = 3 − 12.

The stDNA II of *P.pollicaris* is cleaved by *Alu* I restriction endonuclease as the repeating element contains the AGCT restriction site of this enzyme.

### 6.2.1.3 *G.lateralis* stDNA

The Bermuda land crab *G.lateralis* also contains, together with the dAT satellite, heavy GC-rich stDNA. The repeat length of this stDNA is 2.07 ± 0.1 kb (LaMarca et al. 1981). There are 1.6 · 10⁴ copies of this sequence in each cell. The repeating unit contains a pyrimidine-rich region about 170 bp in length around position 0.64 kb of the *Eco*RI monomer. 20% of the monomer contains a second such region close to 1.8−2.07 kb. The base sequence in positions 495−663, 683−807 of the *Eco*RI monomer has been determined (LaMarca et al. 1981) (Fig. 15). 42% of the bases are adjacent TT and CT dimers.

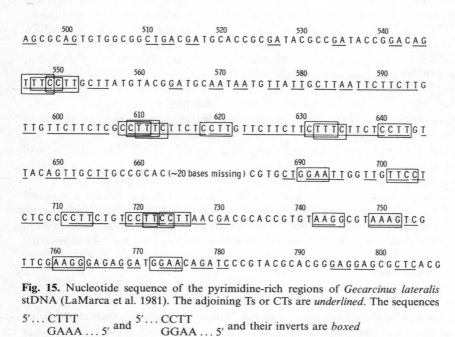

**Fig. 15.** Nucleotide sequence of the pyrimidine-rich regions of *Gecarcinus lateralis* stDNA (LaMarca et al. 1981). The adjoining Ts or CTs are *underlined*. The sequences

$$5'\ldots\frac{CTTT}{GAAA}\ldots 5'\quad\text{and}\quad 5'\ldots\frac{CCTT}{GGAA}\ldots 5'\quad\text{and their inverts are }boxed$$

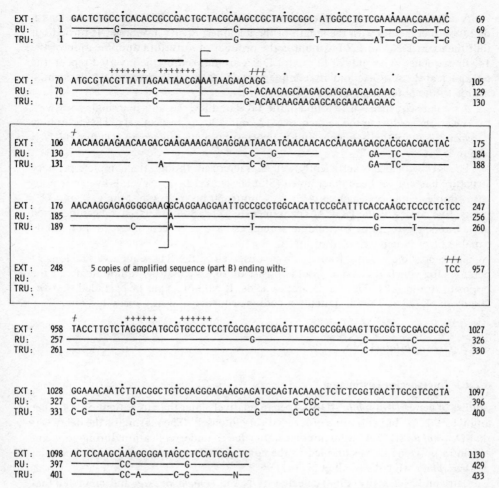

**Fig. 16.** Comparison of the nucleotide sequences of three cloned repeating units of *G.lateralis* stDNA (*EXT, RU, TRU*) (Bonnewell et al. 1983). The amplified *EXT* sequence is in *square brackets*. Denotes the inverted tetranucleotide ends of the amplified region; + indicates other inverted repeats; direct repeats are *boldly underlined*. Omissions correspond to deletions. The sequence rich in AAG/C triplets are *bracketed*

The *Eco*RI monomer is not characterized by internal repetition. It represents a set of several divergent sequences differing both in length and primary structure (Skinner et al. 1983a). The *Eco*RI sequence isolated from different clones, as a rule, differ from each other. The authors do not exclude that all the 16,000 copies in the genome may not be identical.

The *Eco*RI monomer was cloned in the pBR322 plasmid. At an analysis of about 24 clones, two showed a significant macroheterogeneity. One of the clones contained a truncated variant of the *Eco*RI monomer (TRU; 1.674 kb) and the other, on the contrary, and extended one (EXT; 2.639 kb) (Bonnewell et al. 1983).

A detailed analysis was done also of one of the clones (RU) whose insert length of 2.089 kb coincides with the length of the uncloned *Eco*RI monomer. It turned out that the extra DNA in EXT variant is the product of a fivefold amplification of the 142 bp sequence. The ends of this extra DNA are flanked by the inverted repeat, the tetranucleotide AGGA, and its complement (Fig. 16). The extra DNA is located nearer to the 5' terminus of the sequence.

A similar sequence is also found in the TRU and RU variants in a single copy with two deletions, reducing the sequence to 127 and 130 bp, respectively. The homology between the amplified sequence and the corresponding sequence in RU and TRU is about 83%.

The RU sequence is sufficiently well characterized, though the complete primary structure has not yet been determined (Skinner et al. 1983a). Several diverse regions with simple sequences, such as $AG_{n=1-7}$ up to 60 bp in length, long $G_{20}$ and $C_{23}$ tracts, five copies of the CGCAC pentamer, $(C_2T)_{15}$ and others have been detected in it. Interspersed purine/pyrimidine sequences have also been observed, potentially capable of producing Z form of DNA.

Three possible reading frames can be observed in the RU sequence. The longest peptides for which they code would consist of 163, 97, and 49 amino acids; in the opposite strand – 84, 77, and 23 amino acids. It cannot as yet be excluded that the *G.lateralis* 1.723 stDNA is actually a coding sequence and not a satellite one.

## 6.2.2 *Drosophila* stDNA

### 6.2.2.1 Interspecies Distinction

Studies of genus *Drosophila* stDNAs began relatively later than investigations of other animal stDNAs, but with time proceeded very intensely. The reason for the delay was that *Drosophila* stDNAs were not detectable due to underreplication during polytenization in polytene tissues and led to the controversial character of the initial studies of *Drosophila* stDNAs (Gall et al. 1971).

Laird and McCarthy (1968) detected stDNA in *D.melanogaster* and suggested that it was extranuclear DNA due to the variability of its content in different preparations. Rae (1970) also noted the variability of this DNA and assumed that it could be of viral origin. Travaglini et al. (1968) thought the stDNA of *D.melanogaster* to be a mitochondrial DNA as the satellite was absent in the nuclei of salivary glands. Gall et al. (1971) showed that this was due to the presence of polytene cells in the analyzed salivary glands. The stDNAs detected by Gall et al. (1971) in *D.melanogaster* and *D.virilis* constituted 8 and 41% in the diploid tissue total DNA, respectively (Fig. 17).

The number of stDNAs detectable in individual species of the *Drosophila* genus has increased with the development of DNA separation technique using DNA-binding ligands. Thus, two more stDNAs were added to the four known stDNAs of *D.melanogaster* (Barnes et al. 1978), and eight newly detected ones were added to the four known stDNAs of *D.hydei* (Renkawitz 1979).

New satellites are revealed by resolving known stDNAs into constituent components or by detecting stDNAs comprising an extremely small part of a genome (less than 1%) and require specific conditions to be detected. It is not excluded that with

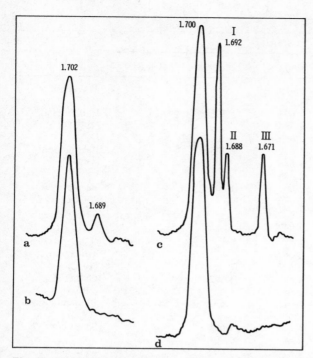

**Fig. 17.** Distribution of *D.melanogaster* and *D.virilis* DNA in the CsCl density gradient (Gall et al. 1971). In both species stDNAs are present in diploid tissues (brain and imaginal disks), but are not detectable in polytene tissues (salivary glands). Imaginal disks and brain of *D.melanogaster* (**a**) and *D.virilis* (**c**); salivary glands of *D.melanogaster* (**b**) and *D.virilis* (**d**)

time new stDNAs will be found in species already studied. In such a situation, the stDNA enumeration used so far for *Drosophila* (stDNA I, stDNA II, etc.) leads to misunderstanding and its use seems inexpedient. The designation of stDNAs by their buoyant density values in the neutral CsCl gradient appears more preferable.

Such a designation is used in recent studies. In a case of twin satellites, additional letter symbols can be used, e.g., 1.686a, 1.686b, etc. Table 6 summarizes the data available on the presence of stDNA in different species of the genus *Drosophila*. For convenience, the stDNAs are conditionally classified under three large groups: the light one (with a buoyant density below $1.690 \, g/cm^3$), of moderate density ($1.690 - 1.710 \, g/cm^3$), and the heavy one (over $1.710 \, g/cm^3$). A coincidence between the buoyant densities of stDNAs of different species does not mean that they are identical molecules.

Some discrepancies in the sources cited should be noted. Thus, Travaglini et al. (1972) reported a 1.711 satellite in *D.virilis*, not mentioned by other authors. Schweber (1974) observed four stDNAs in *D.virilis*, three of which coincide with those found by Gall et al. (1973) and Blumenfeld et al. (1973), while the fourth satellite, 1.704, has not been detected by other authors.

Table 6 shows wide ranges of variation in the number of density stDNAs in the genus *Drosophila*. In some species no stDNAs are revealed. Others contain up to four

**Table 6.** StDNAs of *Drosophila* genus

| Species | Major component | stDNA | | | Source |
|---------|------------------|-------|-------|-------|--------|
|         |                  | Light | Moderate | Heavy | |
| *D.melanogaster* | 1.698 | 1.669<br>1.685 | | | Travaglini et al. (1972) |
|         | 1.701 | 1.673<br>1.686<br>1.688 | 1.705 | | Endow et al. (1975) |
|         | 1.701 | 1.672<br>1.686<br>1.688 | 1.705 | | Brutlag et al. (1977a) |
|         | 1.700<br>1.703<br>1.715 | 1.662<br>1.672<br>1.675<br>1.686<br>1.688 | 1.706 | | Barnes et al. (1978) |
| *D.virilis* | 1.698 | 1.664<br>1.685 | | 1.711 | Travaglini et al. (1972) |
|         | 1.700 | 1.671<br>1.688 | 1.692 | | Gall et al. (1973) |
|         | 1.700 | 1.671<br>1.687 | 1.692 | | Blumenfeld et al. (1973) |
|         | 1.700 | 1.670<br>1.687 | 1.691<br>1.704 | | Schweber (1974) |
|         | 1.700 | 1.671<br>1.687 | 1.692<br>1.693 | | Mullins and Blumenfeld (1979) |
| *D.mauritiana* | 1.700<br>1.705<br>1.715 | 1.663<br>1.673<br>1.686 | 1.693<br>1.694<br>1.696 | 1.734 | Barnes et al. (1978) |
| *D.yakuba* | 1.700<br>1.704<br>1.714 | 1.663<br>1.674<br>1.688 | 1.709 | | Barnes et al. (1978) |
| *D.erecta* | 1.701<br>1.704<br>1.715 | | 1.692<br>1.693 | | Barnes et al. (1978) |
| *D.nasutoides* | 1.702 | 1.665<br>1.669<br>1.682<br>1.687 | | | Cordeiro-Stone and Lee (1976) |
| *D.nasuta* | 1.698 | 1.661<br>1.669<br>1.685<br>(traces) | | | Travaglini et al. (1972) |
| *D.simulans* | 1.698 | 1.669 | 1.692 | | Travaglini et al. (1972) |
|         | 1.700<br>1.704<br>1.715 | 1.673<br>1.676<br>1.686 | 1.693<br>1.694<br>1.695<br>1.707 | 1.734 | Barnes et al. (1978) |

(continued)

**Table 6** (continued)

| Species | Major component | stDNA | | | Source |
|---------|-----------------|-------|---|---|--------|
| | | Light | Moderate | Heavy | |
| D.hydei | 1.698 | 1.685 | 1.692 | 1.711 | Travaglini et al. (1972) |
| | | 1.677 | 1.693 | 1.714 | Renkawitz (1979) |
| | | 1.680 | 1.694 | | |
| | | 1.682 | 1.696a, b | | |
| | | 1.689 | 1.698 | | |
| | | | 1.702 | | |
| | | | 1.704 | | |
| D.montana | 1.700 | | | 1.712 | Gall et al. (1973) |
| D.flavomontana | 1.700 | | | 1.712 | Gall et al. (1973) |
| D.borealis | 1.700 | | | 1.721 | Gall et al. (1973) |
| D.ezoana | 1.700 | | | | Gall et al. (1973) |
| D.littoralis | 1.700 | | | | Gall et al. (1973) |
| D.imeretensis | 1.700 | | | | Gall et al. (1973) |
| D.lacicola | 1.700 | | | | Gall et al. (1973) |
| D.novomexicana | 1.700 | 1.676 | 1.691 | 1.721 | Gall et al. (1973) |
| D.a.americana | 1.700 | 1.676 | 1.691 | 1.721 | Gall et al. (1973) |
| D.a.texana | 1.700 | 1.676 | 1.691 | 1.721 | Gall et al. (1973) |
| D.gymnobasis | 1.698 | | 1.690 | | Miklos and Gill (1981) |
| D.silvarentis | 1.698 | | 1.690 | | Miklos and Gill (1981) |
| D.grimshawi | 1.698 | | 1.690 | | Miklos and Gill (1981) |

stDNAs (*D.virilis, D.nasutoides*) occupying 40–60% of a genome. The number of stDNAs detected in *D.melanogaster* and *D.simulans* is even greater.

### 6.2.2.2 *D.melanogaster* stDNA

Up to the present six stDNAs have been detected in *D.melanogaster* (Barnes et al. 1978). According to Peacock et al. (1978), the total amount of stDNAs constitutes about 25% of a genome. All the six stDNAs are found in approximately equal amounts (3.0–4.7%).

Investigations carried out in Gall's laboratory have established the base sequence in the repeating element of several *D.melanogaster* stDNAs (Endow et al. 1975).

A common formula of the L-chain of *D.melanogaster* stDNAs 1.673, 1.686, and 1.705 was derived:

$$5' \ldots (AAN)_m (AN)_n \ldots 3'.$$

Later Endow (1977) used restriction endonuclease *Mbo*II to draw detailed information on the structure of the 1.705 stDNA of *D.melanogaster*. More than 95% of this DNA is digested by this endonuclease to form fragments shorter than 25 bp. The restriction site of the *Mbo*II, 5' ... GAAGA ... 3', is in the dimeric sequence of 1.705 stDNA, 5' ... AGAAGAGAA ... 3'. Thus, the chemically obtained data on the stDNA sequence in *D.melanogaster* was confirmed by restriction analysis.

    Important information on the structural organization of different stDNAs of
*D.melanogaster* was gained by gene engineering. StDNA cloning showed that individual stDNAs consist of different sets of molecules which as yet are not separable by
any known physical method. The only procedure for such separation is cloning in a
bacterial cell.
    The cloning of simple stDNAs comprising the majority of *D.melanogaster*
stDNAs, is not a simple task as recombinant plasmids are not stable and their cloning
reveals complex recombination processes that have not been fully clarified.
    Brutlag et al. (1977b) constructed hybrid plasmids with tandemly arranged
repeating sequences of *D.melanogaster* stDNAs. Recombinant plasmids containing
long regions of 1.672 and 1.705 stDNAs were unstable. If the construction of recombinant molecules produced hybrid DNAs with a satellite fragment length of $5-10 \cdot$
$10^3$ bp the recombinant plasmids obtained from bacteria contained stDNAs with a
length less than $1.6 \cdot 10^3$ bp.
    The clones containing satellite segments longer than $10^3$ bp were, as a rule, hetero-
genous in size. Although subcloning reduced the heterogeneity of these plasmid
populations, the continuing growth of the cells led to further variations in the size of
the repeating regions. It is assumed that this *rec*A-independent instability of the
tandemly repeating sequences results from irregular intramolecular recombination
occurring in the replicating DNA molecule by a mechanism similar to sister chroma-
tid exchange in eukaryotes.
    The data on the structural organization of *D.melanogaster* stDNAs have been
summarized in papers by Brutlag et al. (1978) and Peacock et al. (1978).
**StDNA 1.672** consists of two types of molecules with a closely related sequence
(Brutlag and Peacock 1979), the structure of which was found by the "nearest neighbor" method and by cRNA analysis. The repeat unit of one type is a pentamer and
that of the other a septamer:

            5'... AATAT
                  TTATA ... 5'        – 60%
            5'... AATATAT
                  TTATATA ... 5'      – 40%.

The molecule consisting of pentamers is more homogeneous; the stDNA chain consisting of septamers is interspersed with GC pairs though with a lower frequency – less
than one GC pair per 50 bp.
    *Hae*III digests stDNA 1.672 forming a heterogeneous set of fragments with a
molecular mass of $0.3-10 \cdot 10^6$ daltons. The sequence digested by *Hae*III is not
included in the basic repeat structures of this satellite (Shen et al. 1976).
**StDNA 1.705**, similar to stDNA 1.672, consists of two types of molecules with closely
related repeats (Fry and Brutlag 1979). The molecule consisting of tandemly arranged
pentamers comprises 85% of all the DNA; the other molecule consists of septamer
segments:

            5'... AAGAG
                  TTCTC ... 5'        – 85%
            5'... AAGAGAG
                  TTCTCTC ... 5'      – 15%.

Fry and Brutlag demonstrated that in all the hybrid plasmids containing one of the 1.705 stDNA sequences the length of the satellite region is 275–1000 bp on the average. This means that these sequences in the long strand of chromosomal DNA are segregated at least by this length.

The structural organization of stDNA 1.705 of *D.melanogaster* was studied earlier by Shen and Hearst (1977) using the photochemical cross-linking reaction with 4,5,8-trimethylpsoralen (trioxalen) followed by electron microscopy. This stDNA is a polypyrimidine: polypurine acid, as follows from the structure of the repeating sequence. Trioxalen cross-links the double DNA strand at the sites where neighboring pyrimidines are located in the opposite chains. Such sites in stDNA 1.705 are detectable regularly at an interval of about 250 bp.

Some results of the study by Brutlag et al. (1978) provide evidence that regions of different stDNAs in vivo can be adjacent. An asymmetry on the "light side" of the peak was found in the 1.705 fraction. Analysis of this fraction showed it to be the result of covalent coupling of components 1.705 and 1.672. Such a picture is observed in preparations of DNA with a length of up to 1500 bp.

**StDNA 1.686**. Its structure can be presented as follows (Brutlag et al. 1978):

$$5' \ldots \text{AATAACATAG}$$
$$\text{TTATTGTATC} \ldots 5' \quad - 80\%$$
$$5' \ldots A_5 T_3\ C_1 G_1$$
$$T_5 A_3 G_1 C_1 \ldots 5' \quad - 10\%$$
$$5' \ldots A_7 T_1\ C_1 G_1$$
$$T_7 A_1 G_1 C_1 \ldots 5' \quad - 10\%$$

**StDNA 1.690**. This satellite is rich in GpC sequences. Its structure has been insufficiently studied.

**StDNA 1.697** is heterogeneous in composition. About 40% of this satellite is rDNA. After reassociation the heterogeneous distribution profile is observed in the density gradient.

**StDNA 1.688** is more complex in structure than the other *D.melanogaster* satellites.

Shen et al. (1976) were the first to show that stDNA 1.688 is partially digested by *Hae*III and *Hinf*I into a fragment of about 350 bp long and multiple oligomers.

Hsieh and Brutlag (1979a) determined the complete nucleotide sequence of the *D.melanogaster* stDNA 1.688 repeat unit. This satellite comprising about 4% of a genome is localized mainly in sex chromosomes and contains a repeat unit of 359 bp length (Fig. 18). The nature of sequence variability in this DNA was determined by restriction analysis and direct sequencing of: (1) individual monomers cloned by hybrid plasmids, (2) a mixture of 15 tandemly arranged monomers from a cloned segment of this satellite DNA, (3) a mixture of monomers isolated by restriction endonuclease cleavage of total 1.688 stDNA.

Both direct sequencing and restriction analysis indicated the considerable variability of some positions in the stDNA. The majority of positions in a sequence are, however, conservative. Sequence analysis of the mixture of 15 tandemly arranged monomers revealed 9 variable positions out of 359. In addition, four changed positions were observed in the total stDNA. The variable positions are irregularly distrib-

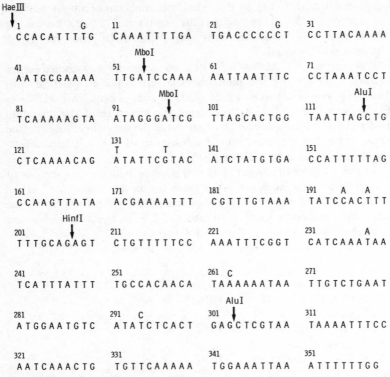

**Fig. 18.** Nucleotide sequence of a 359 bp monomer of stDNA 1.688 from *D.melanogaster* (Hsieh and Brutlag 1979a). Cleavage sites of some restriction endonucleases are indicated

uted in the repeat unit of stDNA 1.688. The bulk of the repeats is homogeneous in sequence.

StDNA 1.688 sequencing experiments indicate that the cytidine residues are not methylated. This includes the CpG-dinucleotides which are frequently modified in other stDNAs.

The most important internal homologies in the repeat unit of *D.melanogaster* stDNA 1.688 (Hsieh and Brutlag 1979a) are shown in Table 7.

On the whole, the monomer sequence does not reveal any internal repeatability thus indicating that it has not been formed from a simple repeating sequence, or that it has diverged to such an extent that it overshadows the existence of a simple precursor.

StDNA 1.688 was shown to include regions with a different distribution of *Hae*III sites. Some of them do not contain *Hae*III sites, while most of the regions contain them at intervals of 359 bp. The long *Hae*III resistant molecules also contain 359 bp tandems and regularly located *Hin*fI sites. At the same time an analysis of the cloned segment of a *Hae*III-resistant DNA revealed a new stDNA 1.688 consisting of a repeat unit 254 bp in length (Carlson and Brutlag 1979). The small fraction of the stDNA 1.688 is comprised of molecules consisting of 254 bp repeats.

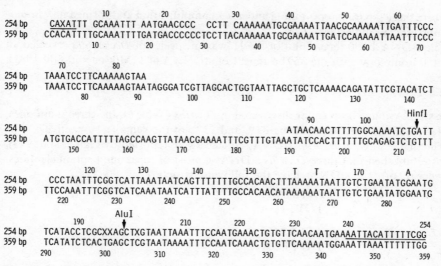

**Fig. 19.** Comparison of nucleotide sequences of 254 and 359 bp repeats of *D.melanogaster* stDNA 1.688 (Carlson and Brutlag 1979). The structure of the undetermined base is marked by the symbol X. *Underlining* indicates that the sequence is uncertain. *Hinf*I and *Alu*I restriction sites in the 254 bp unit are shown. Alternate nucleotides are detected in positions 158, 162 and 179

**Table 7.** Repeating sequence in the *D.melanogaster* stDNA 1.688 cloned monomer (Hsieh and Brutlag 1979a)

| Start position | Repeats |
|---|---|
| 243 | A−T−T−A−T−T−T−G−C−C−A−C−A |
| 351 | A−T−T−T     T−T−T−G−G−C−C−A−C−A |
| 6 | T−T−T−T−G−C−A−A−A−T−T−T |
| 215 | T−T−T−T−C−C−A−A−A−T−T−T |
| 30 | T−C−C−T−T−A−C−A−A−A−A−A |
| 77 | T−C−C−T−T     C−A−A−A−A−A |
| 38 | A−A−A−A−A−T−G−C−G−A−A−A |
| 336 | A−A−A−A−A−T−G     G−A−A−A |
| 57 | C−A−A−A−A−A−T−T−A−A−T |
| 82 | C−A−A−A−A−A−G−T−A−A−T |
| 58 | A−A−A−A−A−T−T−A−A−T−T |
| 262 | A−A−A−A−A−A−T−A−A−T−T |

The 254 and 359 pair repeats are essentially similar to each other with the exception that the 98 bp sequence is absent in the 254 bp unit (Fig. 19). Apart from this sequence, the two monomers are 80% homologous and differ mainly in insertions, deletions, or replacements of single bases. The homology between these two units is indicative of a common precursor of these two types of 1.688 stDNA. An actual separation of these molecules by cloning implies that with the change of the precursor

structure these separate repeat units underwent amplification and formed new homogeneous tandem molecules.

The stDNA 1.690 repeat unit of the Hawaiian species of *Drosophila* revealed an
essential homology with the 359 bp repeat of stDNA 1.688 (Miklos and Gill 1981).

### 6.2.2.3  *D.virilis* stDNA

Four stDNAs have been so far discovered in *D.virilis* (Table 6), the three main ones,
1.671, 1.688, and 1.692 comprising 8, 9, and 23% of the genome, respectively.

The first results on *D.virilis* stDNA sequence were reported by Gall et al. (1973).
It was established that three *D.virilis* stDNAs consist of repeating heptanucleotides.
The repeats of these stDNAs have the following structure:

$$1.692 \qquad 5'\ldots ACAAACT$$
$$TGTTTGA \ldots 5'$$

$$1.688 \qquad 5'\ldots ATAAACT$$
$$TATTTGA \ldots 5'$$

$$1.671 \qquad 5'\ldots ACAAATT$$
$$TGTTTAA \ldots 5'.$$

The buoyant density of the stDNA can be calculated accurately from the structure
of the repeating heptanucleotide. The established sequence clearly indicates the relationship between *D.virilis* stDNAs. It is difficult to attribute stDNAs to one family
just on the grounds of the physical properties of a DNA molecule (buoyant density,
$T_m$). StDNA families can have very close properties if the length of the repeating
element significantly exceeds 7 bp. Any substitution in the case of a short repeat can
lead to essential changes in physical properties of the long molecules. Such a situation
was confirmed by Blumenfeld et al. (1973) who showed that heterologous hybrids
formed by a combination of six separate strands of three *D.virilis* stDNAs and
differing only in one base in the repeating heptanucleotide, melt at a temperature of
$10°-31°$ below that of the homologous complexes. At the same time, the complexes
with two replacements out of seven are unstable at room temperature. The important
point in these investigations is that such "cross-hybridizations" are frequently used
to estimate the relationship of the DNA sequence. It turned out that cross hybridization data can lead to incorrect results and conclusions if the repeating sequences are
very short and differ by at least one base.

The *D.melanogaster* stDNAs do not reveal similarity of primary structure with
that of *D.virilis* stDNAs, its distantly related species. Mullins and Blumenfeld (1979)
detected an additional stDNA in *D.virilis*, composing 0.1% of nDNA. Like the other
stDNAs of *D.virilis*, it is a repeating heptanucleotide and an isomer of stDNAs 1.671
and 1.692. At the same time, its sequence was more related to the stDNA of
*D.melanogaster* and may be an intermediate link between the stDNAs of the two
species. Its buoyant density coincides with that of stDNA I (1.692); it is designated
as Ic (cryptic) (Fig. 20).

StDNA Ic differs from stDNA 1.687 of *D.virilis* by two base replacements, and
from stDNAs 1.671 and 1.692 by three base replacements. It seems probable that
stDNA Ic can be a linking component between the DNAs of *D.melanogaster* and
*D.virilis* which considerably differ in primary structure.

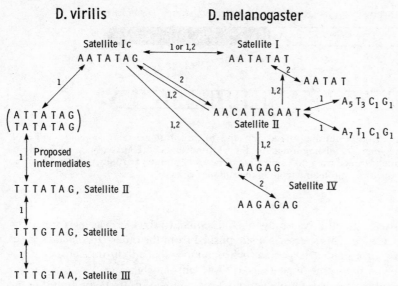

**Fig. 20.** Interrelationship between *Drosophila* stDNAs (Mullins and Blumenfeld 1979). 1 Base substitution, transition or transversion; 2 change of the repeat length

Mullins and Blumenfeld have designed a scheme linking all the known stDNAs of *D.melanogaster* and *D.virilis* (Fig. 20).

### 6.2.2.4 Underreplication of stDNA at Polytenization

It is known that at polytenization of *Drosophila* chromosomes the mitotic hetero-chromatin is either not replicated at all, or is replicated only once or twice. At the same time, the euchromatin regions are replicated 10 to 11 times and yield polytene nuclei with an amount of DNA 1024–2048 times more than that in a diploid cell. StDNA studies permitted to check the theory of heterochromatin underreplication. It has been shown that no stDNAs are detected in the DNA extracted from *D.virilis* or *D.melanogaster* salivary glands and analyzed in the CsCl gradient under normal loading of a cell (Gall et al. 1971). Consequently, stDNAs are not replicated at polytene chromosome formation or, at least, are replicated unproportionally to the remaining part of the genome (see Fig. 17).

Gall et al. (1971) demonstrated that cRNA synthesized on *D.virilis* stDNA 1.671 is hybridized with the centromeric heterochromatin of mitotic chromosomes, though binding with the Y-chromosome is insignificant. The same cRNA binds with α-heterochromatin of the salivary gland nuclei. The degree of bindings with the diploid and polytene cell nuclei is the same, despite their significant difference in DNA content. These data provide evidence that at polytenization the α-heterochromatin is either not replicated at all, or that the replication is insignificant. Polytenization induces the replication either of all, or almost all the hrDNA of euchromatin and β-heterochromatin (Fig. 21).

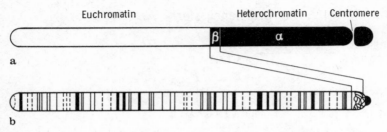

**Fig. 21.** A comparison of mitotic (**a**) and polytene (**b**) chromosomes of *Drosophila* (Gall et al. 1971). Three regions are distinguished in the chromosomes: stDNA-containing α-heterochromatin not replicating during polytenization, β-heterochromatin partially replicating during polytenization, euchromatin mainly containing unique sequences

An exhaustive quantitative study of *D.nasutoides* stDNAs was performed by Cordeiro et al. (1975). This species is distinguished from the other *Drosophila* species by the presence of a pair of large heterochromatic V-shaped chromosomes. In polytene cells this chromosome appears as a "dot" chromosome.

In the diploid tissues of *D.nasutoides* more than 60% of nDNAs are found in the CsCl gradient as four stDNAs, while no stDNAs are detected in polytene cells. These investigations showed that the heterochromatic nature of the V-shaped chromosome is due to the presence of all the four stDNAs in this chromosome. The dot shape of this chromosome in polytene cells is apparently the result of stDNA underreplication during polytenization.

The underreplication of stDNAs in polytene tissues is a common occurrence and has been illustrated with several *Drosophila* species as examples. It has been shown that the underreplication of stDNAs is typical of endopolyploidization (Schweber 1974; Renkawitz-Pohl and Kunz 1975; Endow and Gall 1975). Furthermore, Endow and Gall noted that stDNA underreplication in polyploid tissues is not a general phenomenon. In the highly polyploidized cells of *Dytiscus marginalis* the amount of stDNAs remains at the same level as in diploid cells. Mouse fibroblast polyploid cell DNA also contains a normal amount of stDNAs.

### 6.2.2.5 Chromosomal Location

Chromosomal location of individual stDNAs from different species of the *Drosophila* genus has been the subject of many studies (Gall et al. 1971; Henning 1972; Holmquist 1975; Cordeiro-Stone and Lee 1976; Steinemann 1976; Wollenzien et al. 1977).

In summarizing the results of these studies, the conclusion can be drawn that location of stDNAs is limited by the α-heterochromatin region of chromosomes (Fig. 21). The in situ hybridization technique provides the fine structure of heterochromatic regions for viewing the mutual arrangement of stDNAs and their quantitative distributions.

Peacock et al. (1978) have investigated in detail the location of individual *D.melanogaster* stDNAs in chromosomes. According to these authors, the locations of stDNAs are limited by the pericentric heterochromatin of three autosomes, by the proximal heterochromatin of the X-chromosome, and by the entirely heterochromatic

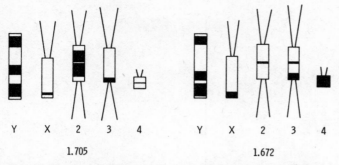

Y   X   2   3   4          Y   X   2   3   4

1.705                          1.672

**Fig. 22.** Cytological hybridization of stDNAs 1.705 and 1.672 with *D.melanogaster* chromosomes (Peacock et al. 1978). Heterochromatin regions are shown as *squares*. Euchromatin is
represented by *lines*

**Table 8.** Distribution of stDNAs in chromosomes (%) of *D.melanogaster* (Peacock et al. 1978)

| stDNAs | Chromosomes | | | | | Content in a genome (%) |
|---|---|---|---|---|---|---|
| | X | Y | 2 | 3 | 4 | |
| 1.672 | 5 | 68 | 1 | 4 | 22 | 4.2 |
| 1.686 | 7 | 50 | 24 | 18 | 1 | 3.3 |
| 1.688 | 48 | 44 | 4 | 3 | 1 | 4.6 |
| 1.705 | 5 | 52 | 33 | 9 | 1 | 4.7 |
| 1.697 | 46 | 38 | 10 | 6 | – | 4.5 |
| 1.690[a] | | | | | | 3.0 |

[a] There are no data on the chromosomal localization of this stDNA

Y-chromosome (Fig. 22). Table 8 shows the quantitative distribution of five of the six
stDNAs in different chromosomes.

The studies led to the following conclusions: (1) each stDNA is localized in a
specific region of chromosomes, (2) all the stDNAs are almost entirely localized in the
heterochromatic regions of chromosomes, (3) individual stDNAs are not limited by
any particular chromosome, however, each chromosome is characterized by a quantitatively and qualitatively unique segmentation of its heterochromatin, (4) the X-
chromosome contains almost no other sequences except those of heterochromatic
DNA and rDNA.

Table 9 presents the location sites of individual stDNAs.

## 6.3 Amphibians

### 6.3.1 *Xenopus laevis* stDNA

About 1% of the *X.laevis* nDNA is organized as tandemly arranged repeats with a
length of about 0.75 kb (Lam and Carroll 1983; Meyerhof et al. 1983; Ackerman
1983). The size of the cloned monomers is 741, 745, and 746 bp (Fig. 23). The homol-

**Table 9.** Composite map of stDNA distribution in the mitotic chromosomes of *D.melanogaster* (Peacock et al. 1978)

| X-Chromosome | | Y-Chromosome | | Chromosome 2 | | Chromosome 3 | | Chromosome 4 |
|---|---|---|---|---|---|---|---|---|
| Centromere region | 1.672<br>1.686<br>1.705 | Short arm of Y | 1.672<br>1.705 | Euchromatin–heterochromatin junction region | 1.697<br>1.705<br>1.672 | Euchromatin–heterochromatin junction region | 1.697 | 1.705[a]<br>1.672[a]<br>1.688[a] |
| Nucleolus organizer region | 1.688<br>1.686<br>1.697 | Nucleolus organizer region | 1.697<br>1.686<br>1.688 | Centromere region | 1.688<br>1.686<br>1.672<br>1.705 | Centromere region | 1.688<br>1.686<br>1.672<br>1.686 | 1.697 |
| Heterochromatin–euchromatin junction region | 1.686 | Centromere region | 1.697<br>1.705<br>1.697<br>1.672<br>1.686 | Heterochromatin–euchromatin junction region | 1.697 | Heterochromatin–euchromatin junction region | 1.705 | 1.705 |
| | | Long arm of Y | 1.705<br>1.688<br>1.672 | | | | | |

[a] No region specified

ogy between them is significant (about 95%). No short repetitive sequences are detected in the basic repeat unit and this stDNA can be assigned to the complex class.

A clone containing tandemly arranged 1037 bp sequences has also been isolated. The difference between the 745 and 1037 bp sequences is in the 287 bp insert implanted in the 745 bp sequence at position 115–116 (Fig. 24). Without the insert the homology between the 745 and 1037 bp repeats is 84.3%.

**Fig. 23.** Nucleotide sequence of the clones 741 bp repeating unit of *Xenopus laevis* stDNA (Lam and Carroll 1983). Endonuclease restriction sites are *boxed*

A 55 bp sequence at the start and end of the 287 bp insert indicates a 75% homology with sequences 1–56 and 57–111/112 of the 745 bp and 1037 bp repeats.

The central 170 bp part of the insert confined between the 55 bp repeats is flanked by a direct repeat 6 bp in length and can be evidence of its insertion by the transposition mechanism.

## 6.4 Mammals

### 6.4.1 Guinea Pig stDNA

Density stDNAs in the guinea pig were first detected by Kit (1961). This species has three stDNAs (Corneo et al. 1968, 1970a) (Fig. 25; Table 10).

**α-stDNA.** The first study to elucidate the base sequence and evolution of guinea pig α-stDNA was done by Southern (1970). One of the methods to determine the DNA structure is the degradation of DNA to pyrimidine oligonucleotides in acidic medium in the presence of diphenylamine and the establishing of their structure. Base sequencing analysis of DNA is considerably facilitated if it can be separated into single chains which is often feasible for stDNAs.

```
        10          20          30          40          50          60          70          80
AAGCTTAAAA  AACAAAAAGC  ATTTGAGAAG  CTGACTAGCG  ACGGCAGCTA  TGGTGCGATT  CTGTCAAAAA  GAAAGTAGAG


        90         100         110         120         130         140         150         160
GTCGCTCTTG  AATGCAGATG  CAGAAGATGG  CACAGATGAA  CCTTCTTGAA  ATTAAGCATT  TGAGAAGCTG  ATTAGCGAAG


       170         180         190         200         210         220         230         240
GAAGTTTTGG  TGCTTTAACA  AATGTGAAAT  ATTGGTTACT  CATTAATGTA  ACCAGAAGAA  AAGGCACCTT  TTCCCTGAAA


       250         260         270         280         290         300         310         320
ATGCACCCCT  AGGACTCACA  CGCTGGGCCC  TTTGGGAAAG  AAACAAGGTG  ATTTGCTGCA  TGCGCCATAG  AACAACCATG


       330         340         350         360         370         380         390         400
TGTGGGTGCT  GATTTTGCTG  TCTCCAAAAT  AGCAGTCGCT  CCTGAATGCA  GCATGAAGAA  AATGCCATCG  GCGAACCGCA


       410         420         430         440         450         460         470         480
CAAATAAAGG  GTTCGTACGG  CTTAAAGAAA  CCTGAAAAGT  CACCCAACCA  TGTTTAGCTC  AGAACCCCAC  GGGCAGAAAC


       490         500         510         520         530         540         550         560
AGGTGCTTTA  ACCATTACAC  CAGAGCAAAC  TAGACCTTTT  GGGAAAGGTT  CCGGATGACT  GTAAGAACTC  ATTGTTCCCT


       570         580         590         600         610         620         630         640
TGTGCCTCTG  CATTGACGGG  CAGGGCCGGA  AGTCCATCTT  CTGCTTAATA  AAGTCAATAC  CAACCTGAGG  CCCTTTCCAA


       650         660         670         680         690         700         710         720
ATAGAGCCTG  CCTGAATAGG  TCAGTCGGTA  GAGCGCAGGG  CTCTGGTCGA  TAACCTNGTC  CAAAAGGTTG  TGGGTTCGAT


       730         740         750         760         770         780         790         800
TCCCACCTCT  GCCGGAAACT  CAATTCCANG  GCAGCCCTNG  TCCGCTGTGC  TTTAACCAAC  TCTCGAGAGT  GATGGAAGGG


       810         820         830         840         850         860         870         880
CAAGTGTCAG  AAGTCCGCCT  TCCATTTGGC  TGCCCCTTTT  TTCAACTGGG  GTTGTCCTTC  TAAGCTACGC  CGGTATGTAC


       890         900         910         920         930         940         950         960
TTCGAATTCA  GGCAGGGCTT  CCCCATCTGA  CACCTTGCCC  TTACCTTTGC  ATTAAATTGG  GTATTTTTTC  CCCAGACAGA


       970         980         990        1000        1010        1020        1030        1040
ACTCTTTTTG  GGGTTTGCCT  AGCATCAGCC  TGAAAATGAC  CTTGGCAAAT  TGCTTAATGA  GGACAGAGTT  CTGTGCC
```

**Fig. 24.** Nucleotide sequence of the cloned 1037 bp repeating unit of *Xenopus laevis* stDNA. The 287 bp *boxed* INS sequence occupies position 117–403. *Bold line* indicates two homologous hexanucleotide sequences (Meyerhof et al. 1983)

Fingerprints of cleavage products of the guinea pig α-stDNA H-chain are represented as three main spots, the TT-spot being the principal one. The number of the pyrimidine oligonucleotides obtained from the L-chain is slightly higher. The main oligonucleotide of the L-chain is CCCT.

On the basis of the analyzed data Southern suggested the structure of the basic repeat unit of α-stDNA:

```
5'...CCCTAA           L-strand
     GGGATT...5'      H-strand
```

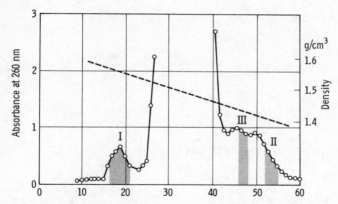

**Fig. 25.** Distribution of guinea pig total DNA in the preparative $Ag^+ - Cs_2SO_4$ density gradient (Altenburger et al. 1977). *Shaded areas* correspond to individual stDNAs

**Table 10.** Buoyant density and % content of guinea pig DNA components (data from Corneo et al. 1968, 1970a)

| DNA component | Buoyant density in CsCl, $g/cm^3$ | % of Total DNA |
|---|---|---|
| Main component | 1.700 | 89.5 |
| α-Satellite (I) | 1.706 | 5.5 |
| β-Satellite (II) | 1.704 | 2.5 |
| γ-Satellite (III) | 1.704 | 2.5 |

The quantitative estimation of all the observable pyrimidine oligonucleotides of the guinea pig α-stDNA shows that about 20% of this satellite is diverged. It is noteworthy that mutations in different positions of the basic sequence occurred with an irregular frequency. The total base composition indicates that the replacement of GC pairs by AT pairs occurred more frequently than reverse replacements. The same conclusion arises from the quantitative estimation of pyrimidine oligonucleotides. Thus, for example, the changes CCCT → TCCT occurred ten times more frequently in the process of evolution than CCCT → CCCC.

To explain the presence of all the observable pyrimidine oligonucleotides, Southern proposed a divergence scheme of the basic sequence, including transitions and transversions as well as insertions and deletions (Fig. 26). On the basis of α-stDNA structural data, Southern developed theoretical concepts on the evolution of this stDNA which were applied later to explain the structural organization of other stDNAs as well. Two variants were considered by Southern.

(1) α-stDNA was formed in one abrupt stage by saltatory replication of the basic sequence (saltatory replication implies a process wherein some small sequence of a genome is replicated ten to hundred thousand times, while the remaining part of the genome is replicated only a few times or not at all; saltatory replications probably occur during one generation). The introduction of $10^6$ mutations into

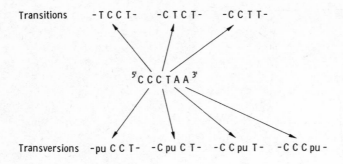

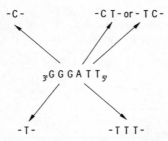

**Fig. 26.** Mutations in the basic sequence of guinea pig α-stDNA (Southern 1970). Other sequences observed in the H- and L-chains can be derived from the basic one by transition and transversion. Some sequences can be obtained by insertions and deletions

the replicated satellite sequence would be required to form the observed α-stDNA structure.

Since the mutation of the satellite sequence proceeded irregularly, it can be assumed that satellite sequence evolution involved selection with preference given to some mutations.

(2) α-stDNA was formed in several stages on the basis of a number of multiplications. If the periods between multiplications were long enough to introduce mutations, the amount of any mutation in the final sequence would be defined by the moment at which mutation took place. If mutation occurred at the initial stages when there were only a few copies of the basic sequence which multiplied later many times, the mutation products must be present with a high frequency in the final sequence. They would be less frequent if mutations occurred at the last stages of stDNA formation.

According to Southern, the second alternative is more acceptable as it gives a quite simple explanation of the currently observed structural organization of guinea pig α-stDNA.

The primary structure of the main repeating sequence of guinea pig α-stDNA excludes the possibility of coding for any protein. Of the six reading frames of this sequence, two give nonsense codons. The other reading frames give repeating dipeptides: proline-asparagine, leucine-threonin, arginine-valine, or glycine-leucine.

Southern, using guinea pig α-stDNA, was the first to show that the length of the stDNA repeat unit determined by reassociation kinetics can considerably exceed the length of the basic repeat. The length of the α-stDNA repeat, derived from reassociation kinetic data without a correction for an imperfect pairing, is $10^5$ bp.

Under restriction endonuclease treatment α-stDNA forms weak bands without a defined regularity (Altenburger et al. 1977). The *Alu*I restriction endonuclease digests stDNA to a significant extent, but the digestion products are not characterized by a regular structure. The authors note that the cleavage pattern is not typical of stDNAs and rather corresponds to the main DNA in heterogeneity. These data are in agreement with reassociation kinetic results according to which the $C_0 t_{1/2}$ of α-stDNA is from 100 to a 1000 times higher than that of other stDNAs (Corneo et al. 1970a).

**β- and γ-stDNA**. Guinea pig β- and γ-stDNAs are related molecules yielding similar pyrimidine oligonucleotides under analysis, but their behavior differs drastically in response to restriction endonuclease digestion (Altenburger et al. 1977). β-stDNA is almost completely resistant to a number of restriction endonucleases thus manifesting a high homogeneity.

γ-stDNA is completely digested by *Bsu*RI restriction endonuclease with the formation of 215 bp long monomers. It should be mentioned that each repeat unit contains two *Bsu*RI restriction sites, the distance between them being one-fourth and three-fourths of the repeat unit length. That is why fragments of intermediate length are found in the cleavage products together with a monomer and its multiple oligomers. The character of distribution along the DNA chain shows that 30–40% of *Bsu*RI restriction sites are lost in the evolution process. The authors presume that the γ-stDNA monomer, consisting of 215 bp, appeared as the result of two consecutive doublings of the quarter-size DNA fragment.

A comparative analysis was carried out on the distribution of total DNA and the DNA from the heterochromatin and euchromatin fractions of guinea pig liver in the $Ag^+ - Cs_2SO_4$ density gradient (Yunis and Yasmineh 1970). The total DNA in the gradient was separated into three components: a heavy stDNA (α-stDNA), the main component, and a light stDNA (β + γ-stDNAs). A fourfold enrichment of the satellite components is observed in the heterochromatin, while the euchromatin is free of stDNAs (Fig. 27).

## 6.4.2 Kangaroo Rat stDNA

Three stDNAs were discovered in kangaroo rat nDNA. StDNAs were first detected in *D.ordii* by Hatch and Mazrimas (1970). Later these authors studied the distribution of total DNAs from 12 kangaroo rat species in the CsCl density gradient and found considerable differences between their stDNA contents (Fig. 28). Table 11 shows the percentage of stDNAs in individual species of the *Dipodomys* genus. The authos used the designations MS for the medium satellite and HS for the heavy satellite. The heavy satellite in the $Ag^+ - Cs_2SO_4$ density gradient separates into two fractions, the α and the β ones (Hatch and Mazrimas 1974). They isolated individual *D.ordii* stDNAs and examined their properties. The stDNAs melt within a narrow temperature range, are characterized by a high renaturation ability, and separate into individual strands in an alkaline gradient.

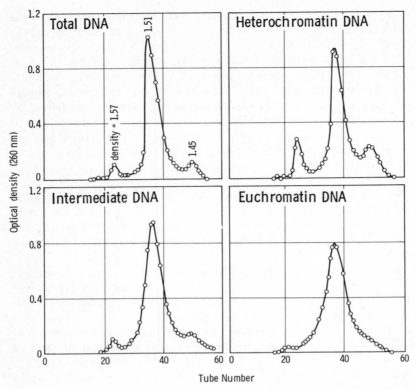

**Fig. 27.** Distribution of guinea pig DNA in the $Ag^+ - Cs_2SO_4$ preparative density gradient (Yunis and Yasmineh 1970)

**Table 11.** Content of DNA components (%) in some species of the genus *Dipodomys* (Mazrimas and Hatch 1972)

| Species | Main component[a] | | Satellite DNAs | |
|---|---|---|---|---|
| | 1.698 | 1.702 | MS (1.707) | HS (1.713) |
| D.ordii | 26 | 23 | 26 | 26 |
| D.microps | 33 | 25 | 19 | 23 |
| D.agilis | 43 | 21 | 23 | 13 |
| D.panamintinus | 44 | 26 | 17 | 13 |
| D.stephensi | 48 | 22 | 20 | 10 |
| D.ingens | 46 | 23 | 16 | 15 |
| D.heermanni | 44 | 23 | 19 | 14 |
| D.merriami | 41 | 28 | 15 | 15 |
| D.elator | 57 | 18 | 15 | 10 |
| D.nitratoides | 41 | 29 | 28 | 2 |
| D.spectabilis | 64 | 21 | 11 | 3 |
| D.deserti | 37 | 44 | 13 | 6 |

[a] Nonsatellite DNA is described by two Gaussian curves

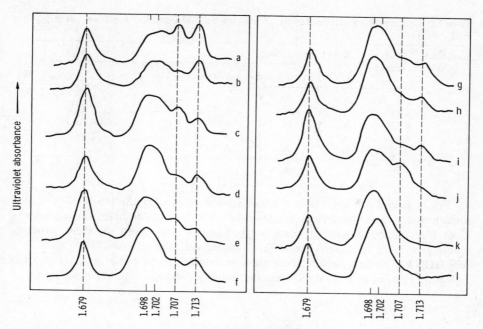

**Fig. 28.** Distribution of total DNAs from 12 kangaroo rat species in the CsCl density gradient (Mazrimas and Hatch 1972). a *D.ordii*; b *D.microps*; c *D.agilis*; d D.panamintinus; e *D.stephensi*; f *D.ingens*; g *D.heermanni*; h *D.merriami*; i *D.elator*; j *D.nitratoides*; k *D.spectabilis*; l *D.deserti*. A dAT polymer ($\varrho = 1.679$ g/cm³) was used as a reference DNA

One of the interesting observations made by the authors is the discrepancy between the GC content values estimated chemically and those calculated from the buoyant density for HS satellites. The GC content is 43.4% and 66.1% in HS-α and HS-β, respectively, whereas their densities are identical (1.713 g/cm³) and correspond to 54.1% of GC. HS-β reveal the highest amount of $m^5C$ observed among the animal DNA fraction analyzed up to the present (6.7% in HS-β and 0.3–1.0% in the other individual fractions).

The primary structure of all the three *D.ordii* stDNA repeat units has been determined (Fry et al. 1973; Salser et al. 1976).

**HS-β stDNA.** The repeat unit of the HS-β stDNA contains ten base pairs and has the following structure (Fry et al. 1973):

$$5' \ldots \text{ACACAGCGGG}$$
$$\text{TGTGTCGCCC} \ldots 5'.$$

Table 12 shows the "mutation" variants of the HS-β stDNA basic 10 bp repeat unit. It follows from Table 12 that the structure of the HS-β stDNA basic sequence and its mutation variants can be schematically shown as

$$-\text{A}-\text{C}-\text{A}-\text{C}-\text{A}-\text{G}-\text{C}-\underbrace{\text{G}-\text{G}-\text{G}}.$$
$$\phantom{-A-C-A-C-A-G}\overset{\nwarrow\;\searrow}{\text{A}\quad\text{G}}\quad\text{G}_4\text{ or }\text{G}_5$$

**Table 12.** Basic repeating sequence of the *D.ordii* HS-β stDNA and its "mutation" variants (Fry et al. 1973)

| Sequence | Relative content | Repeating sequence | | | | | | | | | |
|---|---|---|---|---|---|---|---|---|---|---|---|
| | | 1 | 2 | 3 | 4 | 5 | 6 | 7 | 8 | 9 | 10 |
| 1 | 1 | A | C | A | C | A | G | G | G | G | G |
| 2 | 0.33 | A | C | A | C | A | G | C | | GGGG | |
| 3 | 0.33 | A | C | A | C | A | G | C | | GGGGG | |
| 4 | 0.31 | A | C | A | C | A | G | | | $G_{n,n} = 3-6$ | |
| 5 | 0.15 | A | C | A | C | A | G | A | | $G_{m,m} = 3-5$ | |

Fry et al. did not determine the localization of $m^5C$ in the HS-β stDNA. It is known that cytosine methylation in mammals occurs in the dinucleotide sequence of CpG. If an analogous pattern occurs for the kangaroo rat as well, the above structure would give 7.4% $m^5C$, which is in good accord with the observed $6.7 \pm 1.2\%$.

**MS stDNA.** The MS-satellite sequence was determined by Salser et al. (1976). The repeating element of this stDNA has the following structure (Salser et al. 1976).

$$\begin{array}{ccc} A & \!\!-\!\!A\!\!-\!\! & G \\ \swarrow \searrow & & \downarrow \\ C \quad G & & A \end{array}$$

Due to the simplicity of structure and presence of mutation variations, *D.ordii* MS stDNA is a model suitable for evaluation of different assumptions on the processes of stDNA formation and evolution.

Salser et al. considered two models of the MS stDNA formation process.

(1) MS stDNA resulted from saltatory replication of the AAG triplet. Variant sequences are accumulated after the stDNA chain formation by introduction of mutations. In this case the model must explain why mutations AAG → CAG and AAG → GAG occurred more frequently during stDNA formation than mutations AAG → ACG or AAG → AGG.

(2) The other model of stDNA formation is based on the fact that there was a limited number of mutations in the original short chain containing AAG triplets, followed by multiple saltatory replications of these sequences. On the grounds of this model, the stDNA can be expected to contain CAG triplets spaced at intervals of 6, 9, and 12 nucleotides:

$$1 - (-C-A-G-A-A-G-)_n$$
$$2 - (-C-A-G-A-A-G-A-A-G-)_n$$
$$3 - (-C-A-G-A-A-G-A-A-G-A-A-G-)_n$$

Sequence 1 can be formed from $(AAG)_n$ which contains one AAG → CAG mutation. The formation of sequence 2 from 1 and sequence 3 from 2 can be readily demonstrated. Subsequent recombinations between these sequences will result in the formation of alternating structures, close to the actually observed stDNA structure.

**Table 13.** HS-α stDNA sequences (Fry et al. 1973)

| H-Chain sequence | Substitution in the main sequence | Position | Relative content |
|---|---|---|---|
| | Established sequences | | |
|    1  2  3  4  5  6 | | | |
| 1.  T−T−A−G−G−G | − | | 1.0 |
| 2.  T−T−A−G−A−G | G→A | 5 | 0.5 |
| 3.  T−T−A−G−G−T | G→T | 6 | 0.5 |
| 4.  T−G−A−G−G−G | T→G | 2 | 0.5 |
| 5.  T−T−A−G−G−A | G→A | 6 | 0.3 |
| 6.  T−T−A−G−T−G | G→T | 5 | 0.3 |
| 7.  T−A−A−G−G−G | T→A | 2 | 0.2 |
| | Hypothetical sequences | | |
| 8.  T−T−A−G−T−T | G→T | 5 and 6 | 0.2 |
| 9.  G−A−A−G−G−G | T→G; T→A | 1 and 2 | 0.2 |
| 10. T−T−A−G−G−C | G→C | 6 | 0.2 |
| 11. T−G−A−G−G−A | T→G; G→A | 2 and 6 | 0.1 |
| 12. T−T−A−A−G−G | G→A | 4 | 0.1 |

The second suggested model seems more plausible. It does not need the presence of "hot spots" in the original sequence to explain the specific position of mutations in the stDNAs.

**HS-α stDNA.** Five main sequences have been identified in HS-α stDNA (Salser et al. 1976). It should be noted that sequences 2–5 presented in Table 13 are derived from sequence 1 by substituting one base. Besides the five main sequences, seven additional ones were discovered whose relative content taken together amounts to 1.3 of the content of sequence 1.

The structure of the HS-α stDNA repeat unit was determined from the above data:

$$-G-G-G-T-T-A-$$
$$\quad\downarrow\;\nwarrow\searrow\quad\;\downarrow$$
$$\quad A\;\;A\;\;T\;\;\;G$$

An examination of the *D.ordii* stDNA mutation variants reveals both transitions and transversions occurring in their evolution.

The main sequences of three stDNAs of the kangaroo rat are unrelated, i.e., they have not originated from the same precursor satellite.

Experiments in situ have shown that the *D.ordii* HS-β satellite is localized in centromeric heterochromatin, whereas both HS-α and MS are detected in the chromosome arms exhibiting C-banding (Bostock et al. 1976).

## 6.4.3 Mouse stDNA

Mouse stDNA is one of the most detailed studied stDNAs. As indicated above, it was discovered in 1961 (Kit 1961) (Fig. 29). Corneo et al. (1966) showed this fraction to be of nuclear origin. In the same year it was demonstrated that mouse stDNA consists of relatively short, highly repeating sequences (Waring and Britten 1966). StDNA

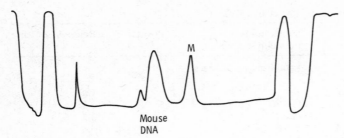

Mouse
DNA

**Fig. 29.** Distribution of mouse brain cell DNA in the CsCl density gradient (Kit 1961). M Reference DNA of *Streptomyces viridochromogenes* ($\varrho = 1.729$ g/cm³)

occupies about 9 % of the mouse genome. Its density (1.690 g/cm³) is lower than that of the main component (1.702 g/cm³).

Later investigations (Flamm et al. 1969; Kurnit et al. 1972; Harbers et al. 1974; Harbers and Spencer 1978; Biro et al. 1975; Shmookler Reis and Biro 1978) were aimed to elucidate its structural features. It was shown that the m⁵C content in stDNA fraction is 1.4 to 3.4 times greater than that of the main component (Bond et al. 1967; Schildkraut and Maio 1968; Salomon et al. 1969).

### 6.4.3.1 Chemical Structure

The complete nucleotide sequence of the mouse stDNA uncloned repeat unit was determined by Hörz and Altenburger (1981) and Manuelidis (1981). The mouse stDNA basic repeat is 234 bp long (Fig. 30). This monomer was obtained by digesting the total mouse stDNA with the *Sau*96I restriction endonuclease. The 234 bp repeat proved to consist of four internal subrepeats, three of which are 58 bp long and the

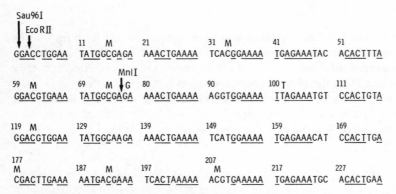

**Fig. 30.** Nucleotide sequence of the A-rich strand of the mouse stDNA repeating unit (Hörz and Altenburger 1981). The sequence of both strands of the *Sau*96I monomer was identified. Endonuclease restriction sites are shown by *arrows*. Homologous nucleotide sequences in the four repeats are *underlined*. The symbol M denotes m⁵C. The CpG sequences at position 208–209 in both strands are only half methylated

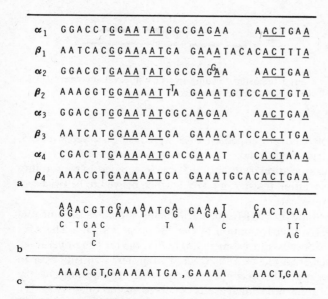

**Fig. 31.** Internal repeats in the mouse stDNA repeating unit (Hörz and Altenburger 1981). **a** The 234 bp monomer is positioned as eight subunits in a manner to visualize the maximum homology between the subunits; identical nucleotides in all the eight repeats are underlined; **b** the consensus sequence derived from eight repeats is shown. Nucleotides occurring once or twice are designated by *small letters*; **c** a possible ancestral sequence of mouse stDNA consisting of three tandemly repeating related nonanucleotides: GA₅TGA, GA₆CT, GA₅CGT

fourth 60 bp long. The G and T insertions at positions 78 and 102 are the only irregularity in the subrepeats. A careful analysis can also distinguish shorter repeating elements within the four subrepeats, corresponding to one-eighth of the 234 bp repeat unit (Fig. 31). Short 28 bp long subrepeats (termed as α-ones) alternate with 30 bp long units (termed as β-ones). The β-unit differs from the α-one by a gap of one nucleotide in the center and by an additional trinucleotide at the 3′-end. If these differences are ignored an initial sequence can be constructed, common for all the eight subrepeats, and consisting of three related nonanucleotides, GA₅TGA, GA₆CT, GA₅CGT (Fig. 31).

In the 234 bp repeat unit 3.4% of the sequence consists of the CpG-dinucleotide wherein almost all the cytosine residues are methylated. In position 208 the cytosine is methylated partially.

## 6.4.3.2 Evolution

Southern (1975) was the first to detect 234 bp long periodical structures in mouse stDNAs using restriction endonuclease *Eco*RII. Electrophoresis of the *Eco*RII endonuclease digestion products revealed a series of bands in the agarose gel. The mobility ratio of the slow and fast bands showed that their length forms an arithmetical series. The intensity of the slow bands decreased with the increase of digestion, as expected in the case of tandemly arranged repeats where the restriction sites are uniformly

1.1/2

R ——— R ———⟍ R ——— R —          R ——— R ———■ R ■ R
■ R ■ R ⟋■ R ■ R     →     ■ R ■ R ■ R ——— R —

1/2

**Fig. 32.** Crossing-over in the "staggered" register, forming two new intervals between stDNA restriction sites (Southern 1975). R denotes a restriction site

spaced. Along with the major bands, intermediate ones were also observed with 0.5, 1.5, 2.5, etc., lengths of the monomer band.

Determination of the reassociation rate and the reassociated duplex stability allowed to calculate the reassociation register of stDNA which proved to be twice less than that of the basic 234 bp repeat.

The presence of intermediate *Eco*RII-formed fragments indicates an unequal crossing-over in mouse stDNA after formation of the 234 bp periodicity. The sizes and relative amounts of the 0.5-mers and 1.5-mers imply that the crossing-over must occur between two molecules arranged in a "staggered" register. Crossing-over in such a register forms two new spaces between the restriction sites. In one molecule the distance is reduced by half, a 1.5 times increase is observed in the other (Fig. 32).

According to Southern, at least four such events occurred at mouse stDNA formation if the repeating sequence was formed as a result of multiplication. The author asserts that it is difficult to imagine a mechanism which would produce shorter repeats within a preexisting repeated sequence and it seems reasonable to draw the conclusion that when more than one periodicity is observed, comparatively long sequences are formed by multiplication of shorter divergent ones. Analysis of the mouse stDNA basic sequence shows a periodicity of 9–18 bp. The following period-icity could form four such units. The periodicity of the *Eco*RII site is 240 bp, while the reassociation register is half of this unit. Thus, according to Southern, four periodicities with 9–18, 36–72, 110–130, and 220–260 bp lengths can reflect the steps of mouse stDNA evolution. It follows from Southern's scheme that an unequal crossing-over took place in stDNA evolution at the last stage after the formation of all four major periodicities (Fig. 33).

It should be mentioned that the determined nucleotide sequence of the mouse stDNA repeat unit, in general, confirmed the principal points in the evolutionary scheme proposed by Southern.

An analysis of long periodical structures of mouse stDNA was performed by Hörz and Zachau (1977) using five restriction endonucleases. All these nucleases produced a major repeat consisting of 234 bp. However, while 80% of stDNA is digested by *Eco*RII, only 5 to 10% is digested by the other four restriction enzymes.

The first type of digestion was termed by the authors as type A, and the second as type B. It follows from these results and the quantitative analysis of joint digestions that the recognition sites of each of the four nucleases are clustered in a separate part of the stDNA. Southern presumed that new sites were inserted into the stDNA as a result of divergency and subsequent distribution by unequal crossing-over.

Hörz and Zachau give an explanation of this phenomenon which differs from that of Southern. They assume that saltatory replication at a defined stage of stDNA evolution was limited by specific regions containing restriction sites of other enzymes

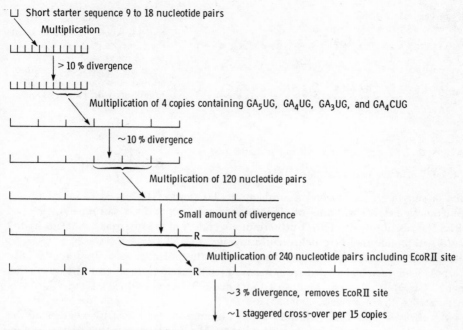

⎵ Short starter sequence 9 to 18 nucleotide pairs

   Multiplication

   > 10 % divergence

   Multiplication of 4 copies containing GA$_5$UG, GA$_4$UG, GA$_3$UG, and GA$_4$CUG

   ~10 % divergence

   Multiplication of 120 nucleotide pairs

   Small amount of divergence

   Multiplication of 240 nucleotide pairs including EcoRII site

   ~3 % divergence, removes EcoRII site

   ~1 staggered cross-over per 15 copies

**Fig. 33.** Evolution of mouse stDNA (Southern 1975). R denotes a restriction site

(excluding *Eco*RII), while the redistribution of these regions, containing areas without restriction sites, occurred by unequal crossing over.

The presence of clusters in mouse stDNA sensitive to some nucleases and differently interpreted by Southern and by Hörz and Zachau, can be the reason for the intermolecular heterogeneity in the stDNA itself. Zardi et al. (1977) demonstrated that mouse stDNA separates into three components in the $Ag^+ - Cs_2SO_4$ density gradient. Satellites I and III are indistinguishable in their melting character and in the density of separate chains in an alkaline gradient. Individual satellite II strands have an identical density in the alkaline gradient. An analysis of separate fractions by restriction endonucleases could clarify the question discussed.

### 6.4.3.3 Minor stDNA

A new stDNA was recently detected in mouse in an amount 10 to 20 times less than of the main stDNA (Pietras et al. 1983). This minor stDNA was found during an analysis of a library of recombinant plasmids containing inserts of mouse DNA repeating fractions. The repeat size of the minor stDNA is 130 bp. It does not exhibit any internal repetition characteristic of the main stDNA. The 94 bp sequence in the minor stDNA repeat has been determined (Fig. 34). The 29 bp segment of the repeat indicates a significant homology with one of the main stDNA subrepeats. The other two-thirds of the minor stDNA do not show an observable homology though some oligonucleotides characteristic of the main stDNA are distinguished.

M. musculus major satellite
sequence nucleotides 205-231.

5'-$G_n$ A C T G A A A A A C A C A T T C G T T G G A A A C G G G A T T T

  G T A G A A C A G T G T A T A T C A A T G A G T T A C A A T G A

  G A A A C A T G G A A A A T G A T G A A A A C C A C A C T $C_n$-3'
  **** ** ******* **** ******
  A A A C G T G A A A A A T G A   G A A A T G C A C A C T

**Fig. 34.** Nucleotide sequence of the cloned repeating unit of mouse minor stDNA. The *first three rows* correspond to minor stDNA, the *fourth* row shows homology with major stDNA (Pietras et al. 1983)

### 6.4.3.4 Chromosomal Location

It was shown by cytological hybridizaton that mouse stDNA is localized in the centromeric heterochromatin of all chromosomes, except the Y-chromosome (Jones 1970; Pardue and Gall 1970). Furthermore, a cRNA in the interphase nucleus hybridized with condensed heterochromatin regions.

Schildkraut and Maio (1968) analyzed the intramolecular distribution of mouse stDNA. The nucleoli isolated from mouse L-cells reveal a three- to fourfold enrichment of stDNA as compared with the total nDNA. One-third of the stDNA is associated with the nucleoli (Fig. 35).

An effort was made by Mattoccia and Comings (1971) to clarify whether all the DNA of mouse heterochromatin consists of stDNA. It was shown that trace amounts of stDNA are present in euchromatin, whereas the nucleolar fraction contained up to 25% of stDNA; extraction of nucleoli with 2 M NaCl increases the share of stDNA in this fraction up to 47%. The buoyant density of euchromatin DNA is equal to $1.702\,g/cm^3$, and that of the main DNA component from the heterochromatin + nucleoli fraction is $1.700\,g/cm^3$. These data indicate that a considerable part of the heterochromatin DNA is nonsatellite DNA.

Brown and Dover (1980) investigated stDNA organization in the mouse X-chromosome. DNA from chinese hamster/mouse hybrid cells containing only mouse X-chromosome was analyzed. The total DNA of hybrid cells was studied with restriction endonucleases which produced both A and B types of digestion. The character of this DNA digestion was compared with that of the total mouse stDNA and some differences were observed. This indicates that each individual chromosome can carry distinct stDNA fractions.

### 6.4.4 Rat stDNA

Rat stDNA was isolated by Pech et al. (1979a) with *Sau*3AI restriction endonuclease. This nuclease digests the total rat DNA into small pieces, leaving the stDNA as a large fragment $10^4$ bp in length. StDNA I comprises 1–3% of the rat genome. The authors determined the base sequence of this stDNA 370 bp long repeating element (Fig. 36).

According to sequence analysis, this repeat consists of alternative 92 and 93 bp units with a close, but not identical sequence.

The authors note four points connected with the structural organization of rat stDNA I:

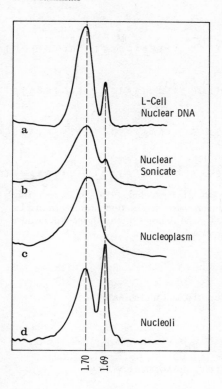

L–Cell
Nuclear DNA

a

Nuclear
Sonicate

b

Nucleoplasm

c

Nucleoli

d

1.70  1.69

**Fig. 35.** Distribution of DNA from mouse L-cell nuclear subfractions in the CsCl density gradient (Schildkraut and Maio 1968). **a** Nuclear DNA; **b** DNA isolated from sonically fragmented nuclei; **c** DNA isolated from the nucleoplasm (nucleoplasm is a supernatant obtained by sedimentation of nucleoli); **d** nucleolar DNA

(1) The homology between individual 92 and 93 bp units is 62–73%. Only two identical regions with a length over 3 bp are observed in all the four units composing the 370 bp basic repeat.

(2) It appears that a minor part of stDNA I has a somewhat different structure as compared with the presented one. Replacements of a defined base are observed at some positions. There are eight such alternative bases. It is not clear whether they reflect only one additional stDNA population with eight replacements in the 370 bp repeat or a set of several molecules with less replacements.

(3) Restriction endonucleases do not digest stDNA I completely. 73% of the *Eco*RI restricts are occupied by monomers, the rest consist of oligomers; in the case of *Hae*III, monomers constitute 91%. This is explained by mutations involving the restrition sites of the mentioned enzymes.

(4) Treatment of the initial stDNA I with *Eco*RI, *Hae*III, or *Hind*III leaves about 15% of the material undigested. However, in each case these 15% are digested with two other endonucleases.

How did the 370 bp sequence originate? According to Pech and co-workers, it was formed from 93 bp unit. Distinctions between the 92 and 93 bp units appeared already within the 370 bp segment. The rat stDNA I observed today was formed from this 370 bp segment by amplification and additional mutations.

The authors believe that at early stages of the amplification process different mutations took place in different copies of the 370 bp sequence; these copies were

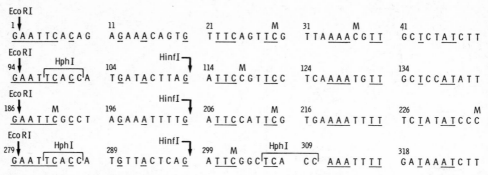

**Fig. 36.** Sequence of the 370 bp repeat unit of rat stDNA I (Pech et al. 1979). Bases occurring at identical positions of four 92 and 93 bp units are *underlined*. Restriction sites are shown by *arrows*. Restriction site *Hph*I is marked with a *bracket*. **M** denotes methylation. Nucleotides encountered in small amounts are given *above the lines*

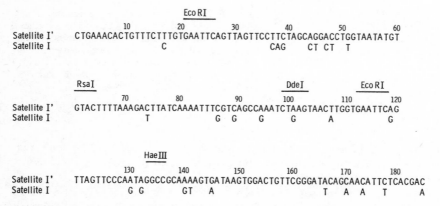

**Fig. 37.** Nucleotide sequence of the 185 bp repeat of rat (*R.rattus*) stDNA I'. In the *second row*, corresponding to stDNA I, only substituted bases are shown (Witney and Furano 1983)

subsequently amplified, forming fractions distinguishable by the presence or absence of certain nucleotides in restriction sites. This may provide an explanation of the segmentary structure of rat stDNA I, corroborated by experimental data indicated in item 4. Incidentally, other stDNAs are also characterized by such a structure. Primary structure determination of the rat stDNA repeating unit in other laboratories (Lapeyre et al. 1980; Sealy et al. 1981) confirmed the data of Pech and co-workers. Witney and Furano (1983) detected a somewhat distinctive stDNA in another rat species, *Rattus rattus*, and called it stDNA I'. The rat satellite I' contains 185 bp tandemly repeated sequences (Fig. 37). This sequence, designated as a' by the authors, is 86% homologous to the 185 bp sequence of stDNA I. They distinguish an "a" sequence in the 370 bp repeat of *R.norvegicus* stDNA I (homologous to the "a'") and "b" sequences. Thus, the structures of stDNA I and stDNA I' are $(a, b)_n$ and $(a')_n$, respectively. Both a and a' are 60% homologous to b.

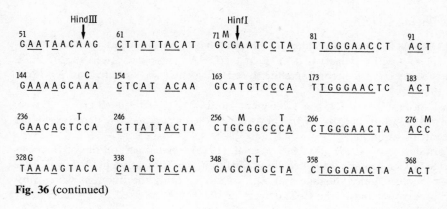

**Fig. 36** (continued)

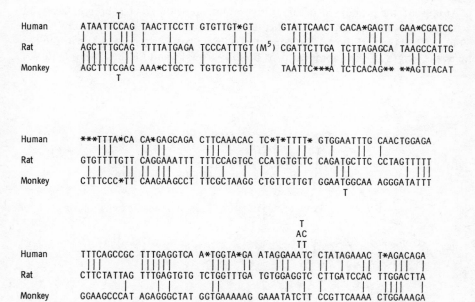

**Fig. 38.** Comparison of the rat DNA 179 bp repeat with the African green monkey α-stDNA and human alphoid stDNA. *Asterisks* denote omissions (Gupta 1983)

*R.norvegicus* contains five times more stDNA I than *R.rattus*. However, the low stDNA I content in *R.rattus* is compensated by stDNA I'. On the whole, the sum of stDNA I and I' in *R.rattus* is close to the amount of stDNA I in *R.norvegicus*.

Digestion of total rat DNA with *Hind*III restriction endonuclease yielded a repeat 179 bp in length (Gupta 1983). A part of this fraction is probably organized in the form of tandem repeats. This repeating sequence is an alphoid one, and exhibits a significant homology with the green monkey α-stDNA (37%) and human alphoid stDNA (41%) (Fig. 38).

### 6.4.5 Bovine stDNA

Eight stDNAs are distinguished in *B.taurus* and have been sufficiently well characterized. The first with a density of 1.713 g/cm³ (stDNA I) was detected in 1962 (Schildkraut et al. 1962). The second satellite ($\varrho = 1.719$ g/cm³) was detected in 1966 (Polli et al. 1966)

Two more stDNAs, III ($\varrho = 1.706$ g/cm³) and IV ($\varrho = 1.709$ g/cm³) (Fig. 39) were found in 1973 (Kurnit et al. 1973). They were isolated by consecutive density gradient ultracentrifugation in $Ag^+ - Cs_2SO_4$ and CsCl.

Cortadas et al. (1977) and Macaya et al. (1978) isolated several minor fractions from the bovine genome using density gradient ultracentrifugation in $Cs_2SO_4$-3.6-bis (acetatomercurymethyl)dioxane and $Ag^+ - Cs_2SO_4$. Only the fractions which produced a narrow band in the CsCl density gradient were referred to as "satellites". Eight such fractions were detected, four of them being already known stDNAs (Table 14).

At present, the primary structure of five stDNA repeat units has been determined. A comparison of these sequences indicates their common origin. Three stDNAs, the 1.706, the 1.711a, and the 1.720b consist of a 23 bp basic sequence, while in the two others the length of the subrepeat is 31 bp (Taparowski and Gerbi 1982b) (Fig. 40).

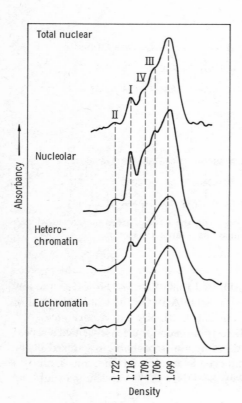

**Fig. 39.** Distribution of nuclear and subnuclear bovine DNA fractions in the CsCl density gradient (Kurnit et al. 1973)

**Table 14.** Content of stDNAs in the bovine genome (Macaya et al. 1978)

| stDNA | % of Total DNA | stDNA | % of Total DNA |
|---|---|---|---|
| 1.706 (III) | 4.22 | 1.715 (I) | 5.07 |
| 1.709 (IV) | 4.57 | 1.720a | 0.14 |
| 1.711a | 1.67 | 1.720b | 0.11 |
| 1.711b | 7.08 | 1.723 (II) | 0.51 |

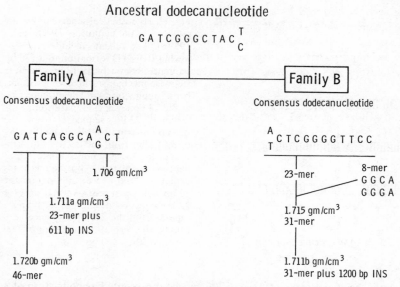

**Fig. 40.** Evolutionary scheme of bovine stDNAs (Taparowsky and Gerbi 1982b)

### 6.4.5.1 StDNA 1.706

Pech et al. (1979b) showed that bovine stDNA 1.706 is characterized by a patchwork structure. The basic repeat of about 2350 bp is comprised of four tandemly connected segments (A-1200, B-247, C-650, D-251). Segments A and C consist of different variants of a 23 bp sequence which, in turn, consists of a dodecanucleotide and a related undecanucleotide (Fig. 41). The undecanucleotide is formed from the dodecanucleotide by a deletion of the eighth or ninth nucleotide. The observable 22 bp sequence is formed from the 23 bp sequence by deletion of the thirteenth nucleotide, usually G.

Segments B and D also consist of a sequence 23 bp in length repeated eleven times. Segment B includes two deletions: 5 bp at one place and 1 bp at the other, segment D has two deletions, each of 1 bp.

Fourteen of the 23 base pairs in the basic repeat of segments A, C, and B, D are identical.

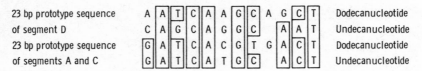

| 23 bp prototype sequence | A | A | T | C | A | A | G | C | A | G | C | T | Dodecanucleotide |
| of segment D | C | A | G | C | A | G | G | C | | A | A | T | Undecanucleotide |
| 23 bp prototype sequence | G | A | T | C | A | C | G | T | G | A | C | T | Dodecanucleotide |
| of segments A and C | G | A | T | C | A | T | G | C | | A | C | T | Undecanucleotide |

**Fig. 41.** Homology of dodeca- and undecanucleotide sequences composing prototype sequence of bovine stDNA 1.706 (Pech et al. 1979b). Identical nucleotides are *boxed*

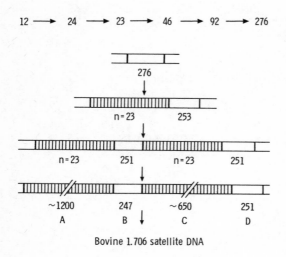

12 ⟶ 24 ⟶ 23 ⟶ 46 ⟶ 92 ⟶ 276

Bovine 1.706 satellite DNA

**Fig. 42.** Hypothetical scheme of bovine stDNA 1.706 evolution (Pech et al. 1979b). The scheme includes the following steps: (1) formation of a 276 bp sequence from a basic 12 bp one by a stepwise mechanism involving a number of duplications and deletion of 1 bp at the 24 bp level; (2) separation of the terminal 23 bp sequence from the 276 bp one and its replication; (3) deletion of 2 bp from the 253 bp, duplication of the 251 bp and its adjacent tandem arrays of the 23 bp sequence; (4) expansion and/or contraction of the 23 bp tandem repeats, formation of a 22 bp sequence, deletion of the 4 bp sequence from the 251 bp sequence and, finally, the formation of about 50,000 copies of the 2350 bp unit

On the whole the structure of stDNA 1.706 can be represented as consisting of a basic 23 bp sequence and its variants derived from it by deletions and mutations at different positions.

The authors have suggested a scheme of stDNA 1.706 formation with the dodecanucleotide as the original structure (Fig. 42).

### 6.4.5.2 StDNA 1.711a

The repeat length of 1.711a is 1413 bp. The primary structure of the uncloned stDNA has been determined (Streeck 1981) (Fig. 43). The sequence can be conditionally divided into three regions. The first 550 bp at the 5′-terminus represent 24.3 tandem repeat variants (21–23 bp) of the stDNA 1.706 prototype sequence. The following, second 611 bp sequence (INS), does not show a repetitive structure. The third region of stDNA 1.711a with a size of 252 bp is close in structure to the D segment of the 1.706 satellite. Their 23 bp prototype sequences differ by 2 bp.

The INS sequence contains open reading frames of over 150 bp and the ATG codon is observed at the start of two frames. This region includes a sequence, complementary to the consensus sequence of the 18S rRNA 3′-terminus and the Hogness box – TATAAATA. It has been suggested that this sequence is suitable for transcription (Streeck 1981).

At the ends of the INS sequence there is an inverted sequence 8 bp in length, seven of which compose a GC pair. At the 5′-terminus the 8 bp sequence is repeated (with a 1 bp deletion). Thus, its structure is reminiscent of sequences observed at the ends of insertion sequences and other transposable elements.

The INS sequence is found also in other fractions of the bovine genome, in particular in stDNA 1.711 b, a part of whose insertion sequence is homologous to the INS sequence of stDNA 1.711 b. The results obtained indicate that the INS sequence is inserted into a molecule close to stDNA 1.706 which probably contained one A or C segment and one D segment. The insertion occurred in the 23 bp repetitive sequence at the point of the deleted 8 bp region.

### 6.4.5.3 StDNA 1.711 b

Its structure has been well studied (Taparowski and Gerbi 1982 b, Streeck 1982). Taparowski and Gerbi found that the first 525 bp and the last 877 bp cloned repeat of the stDNA 1.711 b shared 95 % homology with the 1.715 satellite. An alien sequence is observed in position 526–1712. StDNAs 1.715 and 1.711 b (with the exception of INS sequence) consist of 31 bp repeats. The INS sequence is inserted into one of the 31 subrepeats. Thus, it can be assumed that the 1.711 b satellite originated by amplification of the stDNA 1.715 repeat, modified by insertion of 1200 bp DNA.

The structure of the uncloned INS sequence was detailedly analyzed by Streeck (1982) (Fig. 44a). The INS sequence is flanked by 6 bp direct repeats (CCCTAC). At the 5′-terminus an open reading frame is revealed with a size of 200 bp which is terminated by two TAA codons. At the 3′-terminus several sequences important for transcription regulation are observed. Streeck has shown that the arrangement of the stDNA 1.711 b INS sequence is similar to that of the retrovirus long terminal repeat (LTR) (Fig. 44b).

A comparison of the stDNA 1.711 a and 1.711 b INS sequences indicates a considerable homology of the terminal sequences and a complete unrelatedness of the internal part (Fig. 45).

Streeck has suggested that the INS-sequence of the 1.711 b satellite was the LTR of a large mobile element, possibly of a retrovirus, which was first inserted and then excised by a homologous recombination between two LTRs before being amplified in the stDNA.

### 6.4.5.4 StDNA 1.715 (stDNA I)

Using restriction endonuclease *Eco*RI Botchan (1974) was the first to show that 95 % of the bovine 1.715 satellite consists of the $1400 \pm 50$ bp repeat unit. However, reassociation kinetics showed that the length of the repeat was only 200–250 bp. This means that the 1400 bp unit itself consists of short repeats. Botchan suggested that the stDNA repeat unit has a short substructure and consists of 8 to 20 diverged base pairs, and so reassociation kinetics give an overestimated value.

Restriction endonuclease mapping and fingerprint studies of digested cRNA synthesized on stDNA 1.715 made by other authors did not detect short subrepeats within the 1400 bp repeat; hence it was presumed that stDNA 1.715 belongs to the "complex" class (Roizes 1976; Roizes et al. 1980).

```
GATCACGTGACTCTGCAGGCACT      23

GATCACGTGGCTGATCAAGTCCA      46

GATCACGTGACTGAGCATGCACT      69

GATCACGTGGCT ATCATGCACT      91

GATGACGTGACTGCGCATGCACT     114

GATGACGTGGCTGATCGGGCACT     137

GATCACATGGCTCATCATGCACT     160

GATCACGTGTTT ATCAGGCAAT     182

GATCA GTGACTGA CAGGCGCT     203

GATCATGGGACTGTGCACGCACT     226

GATCACGTGGCTCT CATGCACT     248

GATCACGTGACTGCGCATGCACT     271

GATGACGTGGCTGTTCGGGCACT     294

GATCACGTGTTT ATCAGGCAGT     316

GATCA GTGACTGA CTGGCGCT     337

GATCAGGGGACTGTGCACGCGCT     360

GATCAGGTGGCTGATCAGGAACT     383

GAACACCTGACCGCGCATGCAGT     406

GATCACGTGGCTGGTCGTTCACT     429

GATCACGTGTTT ATCATACGGT     451

GATCACGTGACTGAG AGGCGCT     473

GATCACGTGACTGTGCCGTCAGT     496

GATCACGTGGCTGAGCAGGCACT     519

GATATCGTGACTGAGCATGCACT     542

GATCATGT                    550
```

```
                      G  G
GATCACGTGACTGA TCATGCACT

GATCACGTGACTGATCATGCACT
```

**Fig. 43.** Nucleotide sequence of the bovine stDNA 1.711a repeating unit (Streeck 1981). **a** A *Sau* segment (550 bp). A prototype sequence is presented below. For comparison, a sequence of a prototype repeat unit of stDNA 1.706 is also shown here. **b** An INS sequence. Terminal direct and inverted repeats are denoted by *arrows*. Some sequences are *boxed*: the 991–997, complementary to the 3'-end of the 18S rRNA sequence, the 664–677, the $C_7A_7$, the 1086–1093, the Hogness box. **c** A *Pvu* segment (252 bp). A prototype sequence is given below. For comparison, a prototype sequence of the stDNA 1.706 D segment is also shown here

a        ( 1 . 7 0 6 )

Determination of the *Eco*RI-repeat nucleotide sequence has elucidated controversal points of the 1.715 satellite internal structure. At present, the primary structure has been determined both of the uncloned *Eco*RI-repeat of stDNA 1.715 and of its cloned variants (Gaillard et al. 1981; Pluciennizak et al. 1982; Taparowsky and Gerbi 1982a; Sano and Sager 1982) (Fig. 46).

GCCGGGAGCC  GGGGAGGCAT  TCCACTCTGG  ACAAAGGTCA  TGAGGAAGGA  600

GGCTCGGCAT  ACGCAAATGC  GGGATCGAGC  CTCAGGAGTC  CACCCGGATA  650

TTCTCGAGCA  TCTCCCCCCC  AAAAAAACCG  GAGTCCGCCT  ACTGTATTGC  700

TTTGTGCTCT  CACCTGTGAT  TTCACTGGGG  GCTGTCCCCC  ACCACCATCT  750

CGCTCTCTCT  GTCAAAGATG  TAACTTACAG  CTCCAATTCA  TAAAGTTCCT  800

TGTCATTCTT  CCCTTTAACT  TCCAGCTGAG  TCTCCATCTG  GAGCGCGGAA  850

CCCACCACGC  TTACTAATTA  TGCCTGGGCT  GCTAAGACCC  ACTCGAGAAG  900

GTGTCTAGGG  TGAGGCACCT  TTCGCTATTC  GAGAGGGCGC  CTGCGGCCTA  950

CGTAAGTGGT  GCAAACTTCT  TGTCTTGAAG  TTTGATTGGT  CTTCCGCGTA  1000

AACCAAGCTA  CTCAGTCTCT  TTTCTCCACC  GAATTTTCCT  ACTGAGCTCT  1050

CCTCATACTA  TTATTCTTGA  CATCTCTGAT  TAGCATATAA  ATAGTCGCCT  1100

AGGCCATCTC  TCCTTCGAAT  ACCCTGGATC  AGTTGGGGCT  GGTCCCCGGC  1150

AGGTGGCGAC  C  1161

b

1.  GAACAGGGACCTCAGGAGGCAAT  1184

    ACTCAAAGAGCTGAGCAGACACG  1207

    AACCACGCAA TCAGCAGGCAAT  1229

    AAGCATGGAGCTCAGCAGTGACG  1252

    AATCATGCTGCTCAACTGGCAAT  1275

    AATCAAGCACGTGACCAGGCAGG  1298

    AATCACGCAGCTCAGCTGGCAAT  1321

    TGTCAAGCAGATGAGCCGACAGG  1344

    AATCACGCAGCTCAGCTGGCAAT  1367

10. TGTCATGCAGATGAGCCGGCAGG  1390

    AATCACACAGCTCAGCAGGCCCT  1413

_____

    AATCACGCAGCTCAGCAGGCAAT

    AATCAAGCAGGTCAGCAGGCAAT

c          ( 1 . 7 0 6 )

A comparison of restriction maps of cloned and uncloned stDNA 1.715 has shown that sequence divergence is insignificant, the main difference being in the postsynthetic modification of some restriction sites in the cloned stDNA (Gaillard et al. 1981). A small variability of the repeat length is observed. The length of 60 % of the uncloned repeat is 1399 bp, 40 % of the repeat have 1 bp less. The size of the cloned variant was

```
TGCGTGAGCC GGCATATTGC ATATTGAGTG CATATTGCAT ATTGCATAT  GAGTGCAGCA CTTTCCACAG CATCATCTTT    80
CAGGATCTGG AATAGCTCCA CTGGAATTCT ATCACTGCTG GGAGCCAGTG AGGACTCCG  CCT[ATG]ATAA AGGTTATGAG  160
GAAGGAGGCT CGGCATACGC AAAGGCGGGA TCGAGCTTCA GGAGTCCCCC TGGAAATTCT CGAGCATATA CCCCCAAAAC   240
CAGAGTCTGC CTACTTTCTG CTTGTGCTTT CACCTAAACC TCTGACTTTA CGGGGGGCTC TCCCCCACTA CCTCTCTCTG   320
AAAAAAGAGT TAGCTTACAG TTCCAGT[TAA]TTCCTGG GTGTGACAGT GTTAACCTA  CAAACTCCTT TGGAAATCCT   400
CTAGCCTGCC TGAATAGGTT TTTTCGGCCA CATGGGATTG TTCAGAGCCT CCCAACTGTG AGAGGCAGGA GATGTTCTAA   480
ACTGTCTAAA CACAGATTCT TTTGAGTAGT TACAAGATTG ATTAGAAATT GTATTGGTGA ATGGTTTTTC ACTTGTTGGG   560
CCATTGTTTG CTGCTAAGTT TCCATATCCC TTACCTGCTG TGTCCCTGGC AGTGTATTGA TTAATATAAT TGGTGTAAGT   640
AGTAGCTTTA ATGTTTGTAA CCTGGGACCC TTGAGTTAAT TCTTTTTCTT GTTATAGCCC TGCTCTGTAG TGCTCTGTAG   720
GAATGCAACT TTATCTAATG CTTTTTTGGA GGGTGGCTCC TGACCAACCA CCTTTAGAGA AAAATAAGTT TTCTGAAGAA   800
AAGGTCTTAA AATGTTAACA GGCCTCCGGG CCAGAAGATG ATGCAAATCA CCTAAGCTTT TGCATATGAT AAGTTTGCAG   880
GAAGAAAGCC TGGTTTGCTG CAA[GACTCGA]CCCCTTCCCC CATTATCCTC TATGCATAAC TTAAG[TATA]AAAACTACTT  960
TGAAAAATAA A[GTGCGGGGC]TIGTT[CACCG]TCACCAGTGTC TTACCTTCTC GTTCTTTCTC TTACCTTGTAG GCTGAATTAT 1040
TCAGCCTCTT TTCTCCACTCA AATTCCTCA  CTGAGCTATC TACTCTTAT  ATCCTTAATT AACATTTAAT AACATTTAAT 1120
TAAGCAGTGG TTTCCTGATC CCGTCTCTCC TTCGCCTACG CTGGATCAGC CGGGGCTGGT CCCCGGCA                1198
a
```

```
                 U3              R        U5
LTR    [ TG                  | G   CA |      CA ]
            -75   -23         -20    +15

INS-1.711b [ TG                  | Ḡ   C̄A |      CA ]
               -73   -23         -16    +8
b
```

**Fig. 44. a** Primary structure of the stDNA 1.711b INS sequence (Streeck 1982). 3 bp terminal inverted repeats are shown by *arrows*; the repeating oligonucleotide sequence is *underlined* at the 5'-end; the largest open reading frame is denoted by *boxed* ATG and two consecutive TAA codons. Also *boxed* are signals thought to be important in transcription and polyadenylation. **b** Schematic representation of stDNA 1.711b LTR and INS sequences. *U3* and *U5* correspond to sequences derived from the 3'- and 5'-ends of the viral RNA; R indicates the terminal redundancy. The distances of the regulatory signals from the U3-R boundary (−75, − 23) and R-U5 boundary (− 20, + 15) are given in bp. The *overlined* G and CA in the R-like region correspond to a possible adenylation site, CA, and presumptive cap site, G

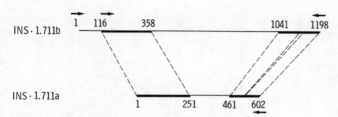

**Fig. 45.** Homology between the INS sequences of stDNA 1.711a and 1.711b (Streeck 1982). Homologous sequences are denoted by *bold lines*. *Arrows* show terminal inverted repeats and their internal repetition in the INS sequence of stDNA 1.711b

1402 bp. The cloned sequence reveals two open reading frames with sizes of 320 and 422 bp, but transcription or translation signals are not distinguished.

Computer analysis indicates that the *Eco*RI repeat of stDNA 1.715 consists of 31 bp subrepeats. About 80% of the *Eco*RI sequence shows more than a 50% homology with the 31 bp consensus sequence. An 855 bp direct repeat (positions 1055–106 and 107–540) is also revealed in the uncloned repeat. The homology between these repeats is 60%. It is assumed that the formation of stDNA 1.715 *Eco*RI repeat proceeded in two stages. First a 950 bp segment was formed from the 31 bp sequence with duplication of the 460 bp region at the second stage (Plucienniczak et al. 1982).

The 31 bp consensus sequence consists of undeca- and dodecanucleotides with eight identical bases and of two tetranucleotides coinciding in three bases (Fig. 47).

The stDNA 1.715 31 bp consensus sequence and the prototype sequence of stDNA 1.720b demonstrate a 55% homology. A 48–55% homology with stDNA 1.706 prototype sequence is also observed. All this indicates their common origin. The variability of sequences in cloned stDNA 1.715 monomers has been analyzed by Roizes and Pages (1982) and Pages and Roizes (1984). It was shown that mutation rates, including deletions and additions, are highly variable along the sequence of repeats. By this criterion, stDNA 1.715 is close to other studied satellites. Analogous also to other stDNAs some endonuclease restriction sites in stDNA 1.715 are clustered in domains, but these domains are essentially overlapping. These indications distinguish stDNA 1.715 from mouse stDNA, where the type B segments are essentially nonoverlapping.

It has been shown that the *Eco*RI repeat of stDNA 1.715 from thymus contains ten times more $m^5C$ than the corresponding fragment from sperm DNA, 0.25 and 2.84 mol%, respectively (Sturm and Taylor 1981). Restriction analysis shows that three *Hpa*II sites in the *Eco*RI fragment from sperm are not methylated completely, whereas in thymus DNA they are methylated fully.

According to Sano and Sager (1982) thymus stDNA 1.715 contains 5.0% $m^5C$, that of liver and brain 4.4% and 2.6%, respectively. Differential methylation in different tissues occurs not at the CCGG sites recognized by *Msp*I and *Hpa*II, but at TCGA as well as other sites (Sano and Sager 1982).

```
   1  GAATTCCCG  TCGTAACTCG  AGAATCCCGC  CGTAACTCGA  GAAAAACCAC  GTGGCTCCCC  CGTCATCGCA  AGATGAAGCC  CTTTCCGCT   ACTGCGCCTC
      CTTAAGGGC  AGCATTGAGC  TCTTAGGGCG  GCATTGAGCT  CTTTTTGGTG  CACCGAGGGG  GCAGTAGCGT  TCTACTTCGG  GAAAGGGCGA  TGACGCGGAG

 101  AGGAGAAGTC  CCACGTTAGG  AATTGGAGGT  CGAAAGGGCC  CTTGGCACCC  TTGATGCGAC  CCACAAAGTT  CCCCGAAATC  CCGGTCTCCC  TCGAGAGGAA
      TCCTCTTCAG  GGTGCAATCC  TTAACCTCCA  GCTTTCCCGG  GAACCGTGGG  AACTACGCTG  GGTGTTTCAA  GGGGCTTTAG  GGCCAGAGGG  AGCTCTCCTT

 201  CACTGAGGTT  TTCCGGCACC  CCCTCCTCTG  AGCCCTTTCT  CCCCTCCTGA  TCTGGACAGG  AGGGTCGACT  CCCCTGCTTT  GTCTGGAAGG  GGTTCCGAT
      GTGACTCCAA  AAGGCCGTGG  GGGAGGAGAC  TCGGGAAAGA  GGGGAGGACT  AGACCTGTCC  TCCCAGCTGA  GGGGACGAAA  CAGACCTTCC  CCAAGGGCTA

 301  CCTTCCGGTC  GCACCTCAGG  ATGAGGCCGG  TCTCACGAAG  ACATTCCAGA  CGTGGCCTCG  TGGGTGTTC   CACATTCCGT  AGGACCCCGA  TTTCCGGTC
      GGAAGGCCAG  CGTGGAGTCC  TACTCCGGCC  AGAGTGCTTC  TGTAAGGTCT  GCACCGGAGC  ACCCACCAAG  GTGTAAGGCA  TCCTGGGGCT  AAAGGGCCAG

 401  CCCTCTTGAT  AAGAACCCGA  TGCCCGGACA  CTCCTCCGAA  CTCCAGCCTG  TGAATGAAGT  CAACACGAAG  GGGCAGTTTT  TCCGTGCATC  GTTCGGAAAA
      GGGAGAACTA  TTCTTGGGCT  ACGGGCCTGT  GGAGAGGCTT  GAGGTCGGAC  ACTTACTTCA  GTTGTGCTTC  CCCGTCAAAA  AGGCACGTAG  CAAGCCTTTT

 501  AACCCAGGT   TCCAAATACA  GCTCGACAAG  CGGCCCTCT   CCCCGGGAC   ATCTCAGAG   GCAAGCGGAG  TTCCATGCCT  CAACCCAAGA  CGAGGCCTGA
      TTGGGGTCCA  AGTTTATGT   CGAGCTGTTC  GCCGGAGAGA  GGGGCCCCTG  TAGAGCTCTC  CGTTCGCCTC  AAGGTACGGA  GTTGGGTTCT  GCTCCCGGACT

 601  CTCTCCTGTC  CCCAGTCTGC  AGGACCCTG   CGATCGGAGT  CTGAAATCAG  AGTACCCTG   CGGTTCCTGC  CTCAACTGGA  GATGAGGCCC  TCTTCAATG
      GAGAGGACAG  GGGTCAGACG  TCCCTGGGAC  GCTAGCCTCA  GACTTTAGTC  TCCATGGGAC  GCCAAGGACG  GAGTTGACCT  CTACTCCGGG  AGAAGGTTAC

 701  CACCAAGCCC  AGTGGAGTCC  CGAGAGGCCC  CTCCCACCTC  CAGTTTCCCT  GGCTTCTCAG  AGCCACCATG  AGAGCCCCC   TGAGGTCACC  TGCACAAGTC
      GTGGTTCGGG  TCACCTCAGG  GCTCTCCGGG  GAGGTGGAG   GTCAAAGGGA  CCGAAGAGTC  TCGGTGGTAC  TCTTCGGGGG  ACTCCAGTGG  ACGTGTTCAG

 801  GAGGGAACCC  AGGGTTTCCT  GCCTCAACCC  GAGAAAGACC  TCAGAGACC   TTCTTCAACA  CGTCTCGAGG  CCACATTCCC  CTACCATGGC  TCGGGAGCAA
      CTCCCTTGGG  TCCCAAAGGA  CGGAGTTGGG  CTCTTTCTGG  AGCTCTCTGG  AAGAAGTTGT  GCAGAGCTCC  GGTGTAAGGG  GATGGTACCG  AGCCCTCGTT

 901  TGACGCGCTC  CCCCTCGCCA  CTCGCATGGA  GACCCGACTT  CCTGGCGCC   CCACGAGAGG  CTCACTGACC  TCGCCGTCGT  ACCTCGTGAG  AAACCGCACA
      ACTGCGCGAG  GGGGAGCGGT  GAGCGTACCT  CTGGGCTGAA  GGGACCGCGG  GGTGCTCTCC  GAGTGACTGG  AGCGGCAGCA  TGGAGCACTC  TTTGGCGTGT

1001  CTGGGGCCGC  CGCTCGAGAA  CAACCCCGAG  ATTCCCCGT   CATCGAGAGA  TGAGGGCCTT  GTGGCCTAGA  CGTCCTCTGC  GACCAATCTC  GGGACTCTC
      GACCCCGGCG  GCGAGCTCTT  GTTGGGGCTC  TAAGGGGCA   GTAGCTCTCT  ACTCCCGGAA  CACCGGATCT  GCAGGAGACG  CTGGTTAGAG  CGCTGGAGAG

1101  TCCAAACGCC  TCAGGAGGCT  TGACTCCCTT  GAGTCACCCT  AAGAGATACC  CGTGCGATT   CGAGAGCAGA  CGCGACGTTC  GCGGACGTTC  TTTGCTTCCA
      AGGTTTGCGG  AGTCCTCCGA  ACTGAGGGAA  CTCAGGTGGG  TTCTCTATGG  GCAGCGCTAA  GCTCTCGTCT  GCGCTGCAAG  CGCCTGCAAG  AAACGAAGGT

1201  CTCGAGATGA  ATGCCTGTCT  CCCCGGGTGC  GTCTGGAATG  CAACCCCAG   ATCCCTGTCG  CCCCTGAGA   GGAACACTGG  CTTCTGACA   CAAGCCTAGA
      GAGCTCTACT  TACGGACAGA  GGGGCCCACG  CAGACCTTAC  GTTGGGGCTC  TAGGGACAGC  GGGGACTCT   CCTTGTGACC  GAAGACTGT   GTTCGGATCT

1301  TGAGGTCTAT  GTCACTCGAG  AGCAATCCCC  AGCTTTCCTT  CGCAACTCGA  ATGGAAGATT  GGGCCAACAC  GACTTGCCT   GGGCCAACAC  AAGAGGCAGC
      ACTCCAGATA  CAGTGAGCTC  TCGTTAGGGG  TCGAAAGGAA  GCGTTGAGCT  TACCTTCTAA  CCCGGTTGTG  CCTGAACGGA  CCCGGTTGTG  TTCTCCGTCG

1401  CT
      GA
```

**Fig. 46.** Nucleotide sequence of the bovine stDNA 1.715 cloned monomer (Gaillard et al. 1981)

|  |  |  |
|---|---|---|
| 2 | ATTCAGGCTGCCTCTTGTGTTGGCCCAGGCA | 21 |
| 34 | GTCCAATCTTCCATTCGAGTTGCGAAGGAAA | 17 |
| 65 | GCTGGGGATTGCTCTCGAGTGACTGCAGGGC | 23 |
| 97 | AATAGACC TCATCTAGGCTTGTGTCCAGAA | 16 |
| 126 | GCCAATGTTCCTCTCCAGGGGCGACAGGGA | 21 |
| 157 | TCTCGGGGTTGCATTCCAGACGCACCCGGGG | 20 |
| 189 | GACAGGCATTCATCTCGAGTGGAAGCAAAGA | 19 |
| 220 | ACCCCGCTCTGCTCTCGAATTGTGACGGGTA | 20 |
| ... | | |
| 282 | AGTCAAGCCTCCTGAGGCGTTTGGAGAGAGG | 16 |
| 313 | TCGCGAGATTGGTCTCTAGGCCATGCAGGAG | 17 |
| ... | | |
| 376 | TCTCGGGGTTGTTCTCGAGCGGCGGCC CCA | 22 |
| 406 | GTGTGCGGTTTCTCACGAGGTACAACGGCGA | 18 |
| 437 | GGTCAGTGAGCCTCTCGTGGGGCGCCAGGGA | 21 |
| 468 | AGTCGGGTCTCCATGCGAGTGGCGAGGGGGA | 21 |
| 499 | GCGCGTCATTGCTCCCGAGCCATGGTAGGGG | 18 |
| 530 | AATCTGGCCTCGAGACGTGTTGAAGAAGGTC | 17 |
| 561 | TCTCGAGGGCTTTCCCGGGTTGAGGCAGGAA | 21 |
| 592 | ACCCTGGGTTCC CTCGACTTGTGCAGGTGA | 22 |
| 622 | CCTCAGGGGGCTTCTCACGGTGGCTCTGAGA | 18 |
| ... | | |
| 682 | TCTTGGGACTCCACTGGGCTTGGTGCATTGG | 18 |
| 713 | AAGAGGGCCTCATCTCCAGTGGAGGCA GGA | 22 |
| 743 | ACCGCAGGTACCTCT GATTTCAGACTCCGA | 18 |
| 773 | TCGCAGGGTCCCTGCAGACTGGGGACAGGAG | 18 |
| 804 | AGTCAGGCCTCGTCTTGGGTTGAGGCATGGA | 22 |
| 835 | ACTCCGCTTGCCTCTCGAGATGTCCCCGGGG | 21 |
| ... | | |
| 896 | ACCTGGGGTTTTTTCCGA ACGATGCACGGA | 19 |
| ... | | |
| 983 | GCATCGGGTTCTTATCAAGAGGGGACCGGGA | 19 |
| ... | | |
| 1073 | AGACCGGCCTCATCCTGAGGTGCGACCGGAA | 19 |
| 1104 | GGTCGGGAACCCCTTCCAGACAAAGCAGGGG | 17 |
| 1135 | AGTCGACCCTCCTGTCCAGATCAGGAGGGGA | 19 |
| .... | | |
| 1197 | CCTCAGTGTTCCTCTCGAG GGAGACCGGGA | 23 |
| 1227 | TTTCGGGGAACTTTGTGGGTCGCATCAAGGG | 17 |
| .... | | |
| 1289 | ACGTGGGACTTCTCCTGAGGCGCTGTAGCCC | 17 |
| 1320 | CAAAGGGCTTCATCTTGCGATGACGGGGGAG | 16 |
| 1351 | CCACGTGGTTTTTCTCGAGTTACGGC GGGA | 24 |
| 1381 | TTCTCAAGTTGCGACGGGGA | 16 |

| Average sequence | ACTCGGGGTTCCTCTCGAGTTGCGGCAGGGA | |
|---|---|---|
| a | b | c |

Fig. 47. Arrays of 31 bp related sequences composing 78% of the 1399 bp repeat of bovine stDNA 1.715 (Plucienniczak et al. 1982). Each 31 bp repeat exhibits no less than a 50% homology with the average sequence. a Initial position of each 31 bp sequence shown in b. c Number of nucleotides occurring at corresponding positions of the 31 bp and average sequence

### 6.4.5.5 StDNA 1.720 b

Pöschl and Streeck (1980) defined the nucleotide sequence of uncloned stDNA 1.720 b. This satellite consists of tandemly arranged 46 bp units (Fig. 48 a). The reported sequence corresponds to the most frequently occurring nucleotides at each position of the uncloned stDNA. The 46 bp unit apparently originated by duplication of a 23 bp fragment. There is a 61% homology between two 23 bp subrepeats of stDNA 1.720 b.

One of the several palindromes in the prototype sequence of stDNA 1.720 b is worthy of mention. It is a large discontinuous palindrome, including nucleotides 6–18 and 31–43 in the F-strand and 4–16 and 29–41 in the S-strand (the F-strand is a fast

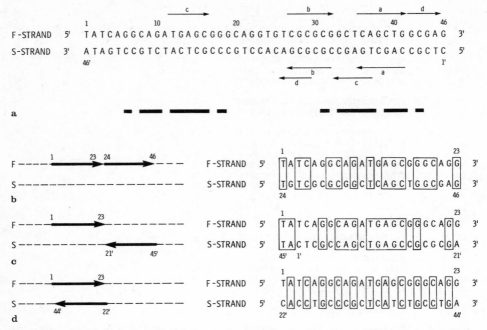

**Fig. 48.** Nucleotide sequence of stDNA 1.720b and the symmetry elements (Pöschl and Streeck 1980). **a** Palindromes (a–c) are denoted by *arrows*, *bold lines* show the large interrupted palindrome; **b** the homology between direct repeats composes 61%; **c** the homology between inverted repeats with the symmetry center between nucleotides 24 and 25 of the F-strand (corresponding to positions 23′ and 22′ in the S-strand), 57% homology; **d** the homology between inverted repeats with the symmetry center at position 13 in the F-strand (corresponding to position 34′ in the S-strand), 48% homology

migrating one in polyacrylamide gel, while the S-strand is a slowly migrating one). A detailed analysis of the stDNA 1.720b structure (Fig. 48b) indicates that it is an imperfect palindrome.

As mentioned above, the two types of stDNA 1.706 segments contain different variants of the 23 bp repeat. A comparison of prototype sequences of these segments and the repeat unit of stDNA 1.720b revealed a considerable homology between them. This gives grounds to assume that stDNAs 1.706 and 1.720b originated from a common sequence. Pöschl and Streeck presume that the evolutionary pathways of these two stDNAs diverged at the 23 bp stage as a greater similarity is observed between the 46 bp prototype sequences of these two stDNAs than between the 23 bp subrepeats of stDNA 1.720b.

### 6.4.5.6 Chromosomal Location

Kurnit et al. (1973) studied the presence of calf stDNA in sex chromosomes. The calf karyotype consists of 58 pairs of acrocentric autosomes and of two readily distinguishable metacentric sex chromosomes. There is no, or very little, heterochromatin in calf sex chromosomes. Calf stDNAs are predominantly located in centromeres of the autosomes. There are few, if any, stDNAs on the X- and Y-sex chromosomes. The C-banding technique reveals constitutive heterochromatin in the centromeres of all the autosomes, but not on the X- and Y-chromosomes.

StDNA 1.716 (I) is encountered in the centromeres of all autosomes. Two-thirds of stDNA 1.722 (II) are observable in centromeres of the autosomes and one-third in the interstitial or telomeric regions. Satellites 1.706 (III) and 1.709 (IV) are localized in the centromeres of most, but not all, autosomes. All four stDNAs in the interphase nucleus are arranged as a cluster. A significant enrichment of the nucleolar fraction by these satellites is observed (Fig. 39). Enrichment in extranucleolar heterochromatin is absent and this, according to the authors, may reflect stDNA association with the perinucleolar heterochromatin.

## 6.5 Primates

### 6.5.1 African Green Monkey stDNA

The DNA of the African green monkey (AGM) consists of four satellite components (Table 15).

The main, α-component is a hidden stDNA and represents 20−25% of the genome. It is detectable in all the tissues and cell lines of the monkey. The α-component is a typical stDNA: it reassociates rapidly, is localized in the pericentromere and replicates later than the main component. The α-component is enriched in DNA preparations isolated from nucleoli. It is assumed that this satellite in the interphase cell is associated with perinucleolar heterochromatin.

Total AGM DNA also displays a heavy component constituting a small variable part of the monkey genome. This component consists of two minor stDNAs designated β and γ.

Hybridization in situ has shown that component γ is localized in the pericentromeric heterochromatin, whereas component β is scattered randomly in the chromosomes. A more intense accumulation is noted in the centromere; at the same time it is revealed also in the chromosome arms. Preparations of the γ-component contain a small amount of a heavier fraction δ with a density of $1.729 \text{ g/cm}^3$ (Kurnit and Maio 1974).

**Table 15.** Components of African green monkey DNA (after data of Kurnit and Maio 1974)

| DNA component | Buoyant density, $\text{g/cm}^3$ | Percentage |
|---|---|---|
| α | 1.699 | 20−25 |
| β | 1.711 | < 1 |
| γ | 1.712 | < 1 |
| δ | 1.729 | ≪ 1 |
| Main | 1.699 | 70−75 |

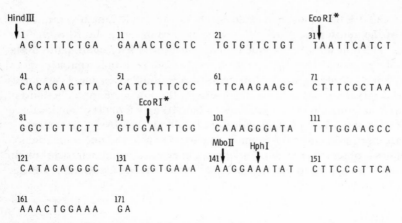

**Fig. 49**. Nucleotide sequence of the AGM *Hind*III monomer (Rosenberg et al. 1978). Inverted repeats (dyad symmetry ≥ 5 bp) are detected at positions 4–8 and 10–14, 68–73 and 78–84, 117–121 and 127–131, 120–125 and 130–135, 107–113 and 145–151. True palindromes (> 7 bp) occur at positions 140–147 and 158–165

Treatment of the AGM total DNA with restriction endonuclease *Hind*III produces a 172 bp long fragment and its multiple oligomers. Purified stDNA also yields a *Hind*III-monomer and its multimer. 77% of the α-stDNA is detected as a *Hind*III-monomer. The monomer represents a set of related, but distinct sequences (Rosenberg et al. 1978).

The primary structure of the uncloned *Hind*III-monomer, i.e., its consensus sequence, is shown in Fig. 49. The *Hind*III-monomer does not contain internal repeats of relatively short oligonucleotide segments. In this sense, α-stDNA represents a complex stDNA.

The *Hind*III-monomer is characterized by an asymmetric distribution of purine and pyrimidine residues in individual chains. Purines or pyrimidines are observed in stretches of 6 to 12 bases. Regions of dyad symmetry and several true palindromes are also encountered.

Analysis of cRNA synthesized on a *Hind*III-monomer revealed a number of minor oligonucleotides composing a rather significant portion of the total amount (2–5%), indicating a defined divergence of the 172 bp repeat sequences.

Later the same group of researchers cloned individual monomer and dimer units of the AGM α-stDNA (Thayer et al. 1981). The nucleotide sequence of four cloned monomers and two dimers was defined. Analysis of the results demonstrated that α-stDNA consists of a set of related, though slightly diverged sequences. Two neighboring monomers in a dimer are no more homologous than randomly isolated monomers.

Part of the α-stDNA is digested with *Eco*RI and *Hae*III or with the *Bsu*RI isoschizomer, 25 and 10%, respectively (Fittler 1977; Graf et al. 1979). These endonuclease sites are clustered in defined stDNA segments and are not distributed evenly. A similar situation was observed also by other authors for mouse stDNA and guinea pig γ-stDNA, where the restriction sites of some endonucleases are also clustered in a part of the stDNA (Hörz and Zachau 1977; Altenburger et al. 1977).

A series of studies by Singer and co-workers was devoted to determining the character of sequences at junction sites of α-stDNA to other sequences (McCutchan et al. 1982; Maresca and Singer 1983; Grimaldi and Singer 1983). Smith's theory on the origin of tandem repeat chains as a result of unequal crossing-over events predicts that the central region of tandem repeats must be more homogeneous than the peripheral one of these repeats, as the latter participate in a less number of crossing-overs and their DNA sequence will undergo a greater divergence (Smith 1976).

As indicated above, only a small number of α-stDNA repeats contain *Eco*RI sites. It is most likely that these sites are located nearer to the ends of the long stDNA tandem chain. Proceeding from this, the following steps were used to isolate the sequence adjacent to α-stDNA. AGM total DNA was preliminary subjected to limited *Eco*RI hydrolysis, 12–16 kb fractions were isolated in the sucrose gradient and inserted into λ *Charon* 4A. Most of the clones, selected by virtue of the presence of α-stDNA, contained both satellite and nonsatellite sequences. The satellite sequence at the junction is mainly characterized by a more extensive divergence as compared with the consensus sequence, which is in good agreement with Smith's theory. Thus, for example, *Hind*III digestion of the cloned α-stDNA sequence, in contrast to the uncloned one, yielded more oligomers than monomers. It has also been shown that the stDNA 172 bp sequence at the junction in one of the clones has a greater homology with the baboon and bonnet monkey stDNA than with its own consensus sequence.

Singer and coauthors have made a detailed study of nonsatellite DNA at the junction with α-stDNA. Three clones were isolated where α-stDNA is linked with different members of the *Kpn*I family. In two cases the size of the nonsatellite segment was more than 6 kb and had one copy of the Alu sequence within it. In the third case

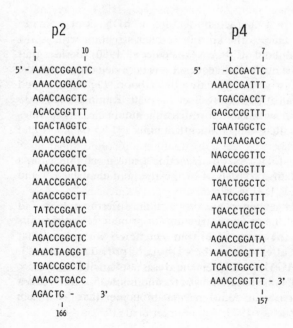

**Fig. 50.** Nucleotide sequence of two cloned monomers (p2 and p4) of AGM deca-stDNA (Maresca and Singer 1983). The frequency of the consensus base at each residue is shown

the nonsatellite sequence is flanked by α-stDNA at both ends which provides evidence that this region is inserted in the satellite sequence (Grimaldi and Singer 1983).

The clone with an inserted short 829 bp nonsatellite segment into the stDNA chain has also been examined (Thayer and Singer 1983). The sequence of this insert is homologous to one of the 6 kb members of the *Kpn*I family called *Kpn*I-RET by the authors. It is flanked by direct repeats 14 bp in length and is encountered only once in the stDNA consensus sequence. It is evident that *Kpn*I-RET was inserted into the satellite chain as a mobile element after the formation of stDNA by amplification.

Besides members of the *Kpn*I family, a previously undescribed sequence consisting of a chain of tandem repeats 10 bp in length was detected in three other clones at the junction with stDNA. The length of the tandem chain in the clones was 9.5–14.7 kb. This sequence was designated by the authors as deca-stDNA. The deca-stDNA structure (the 166 bp and 157 bp long sequence in two clones was determined) indicates that the structure of the tandems essentially diverges from the consensus structure though the tetranucleotide part of CCGG is conserved to a significant degree (Fig. 50).

The deca-satellite sequence is not observed in human or mouse genomes. A considerable difference is observd between the genomes of individual green monkeys in the arrangement of the deca-satellite sequence. Tissue specificity in the same individual is not detected by deca-stDNA.

## 6.5.2 Baboon and Bonnet Monkey stDNA

### 6.5.2.1 Nucleotide Sequence

The 170 bp repeating sequences are a charactistic feature of stDNAs of primates (Donehower and Gillespie 1979). Due to the kinship of their structure with AGM α-stDNA they are designated as alphoid stDNAs (Musich et al. 1980). Besides that of the human and AGM species, the nucleotide sequence has been determined for the repeat units of alphoid stDNAs of two other primates, the baboon (*Papio papio*) and the bonnet monkey (*Macaca radiata*) (Donehower et al. 1980; Rubin et al. 1980).

Digestion of baboon total DNA with *Bam*HI restriction endonuclease generates a 343 bp fragment consisting of related, but nonidentical units 171 bp (*papio* 1) and 172 bp (*papio* 2) in length (Fig. 51).

By digestion of bonnet monkey total DNA by *Hae*III restriction endonuclease, a 342 bp fragment is produced which also consists of two related, but identical 171 and 172 bp sequences (monomers I and II) (Fig. 52).

The baboon and monkey dimer sequences are very similar, differing only in 5 of the 340 positions. A comparison of the primary structures of primate repeating units of alphoid stDNAs indicates that they originated from a common precursor. What is this initial sequence like? To what extent did the stDNAs of primates diverge?

The *Hind*III-monomer of the AGM α-stDNA is the closest to the initial sequence of primate stDNAs. Individual subrepeats of the baboon and bonnet monkey dimer units exhibit a higher homology with the AGM *Hind*III-monomer than with each other (Rubin et al. 1980) (Table 16).

|  |  |
|---|---|
| Papio 1 | AGCTTTCTGAGAAACTGCTTAGTGTTCTGTTAATTCATCTCACAGAGTTACATCTGTATTCGTGGATCTCTTTGCTAGCCTTAT |
| Papio 2 | AGCTTTCTGAGAAACTTCTTTGTTCTGTGAAATCATCTCACAGAGTTACAGCTTCCCCTCAAGAAGCCTTTCGCTAAGACAG |
| Cerco-pithecus | AGCTTTCTGAGAAACTGCTCGTGTTCTGTTAATTCATCTCACAGAGTTACATCTTCCCTTCAAGAAGCCTTTCGCTAAGGCTG |
| Human | ATAATTCTCAGTAACTTCCTTGTGTGTGTATTCAACTCACAGAGTGAACCATCCTTTACACAGAGGAGACTTCAAACACTC |

| Papio 1 | TTCT GTGGAATCTGAGAACAGATATTCGGATCCTTTGAGACTATAGGGCCAAAGGAAATATCCTCCGATAACAAAGAGAAAGA |
| Papio 2 | TTCTTGTGGAATTGGCAAAGTGATATTTGGAAGCCATAGAGGGCTATGGTGAAAAAGGAAATATCCTCAGATGAAATCTGGAAAGA |
| Cerco-pithecus | TTCTTGTGGAATTGGCAAAGGGATATTTGGAAGCCCATAGAGGGCTATGGTGAAAAAGGAAATATCTCCGTTCAAAACTGGAAAGA |
| Human | TTTTTGTGGAATTGCCAACTGGAGATTTCAG CCGCTTTGAGGTCAATGGGAAGTGAGTAGAATAGGAAATATCTCCTATAGAAACTAGACAGA |

**Fig. 51.** Comparison of primate alphoid stDNA nucleotide sequences (two subrepeats of baboon stDNA, papio 1 and papio 2, the *Hind*III fragment of AGM α-stDNA and the 171 bp sequence of human alphoid stDNA (Donehower et al. 1980). Regions of homology are enclosed by a *continuous bold line. Arabic figures* show the boundaries between homologous and nonhomologous areas of papio 1 and man. The short nucleotide sequence 5'... AAGG  and its invert 5'... GGAA  are *underlined*
TTCC....5'      CCTT....5'

**Table 16.** Homology of bonnet monkey, human, and AGM alphoid stDNAs (number of common bases) (Rubin et al. 1980)

| Bonnet monkey monomer | 171 bp *Eco*RI fragment of human DNA | *Hind*III fragment of AGM α-stDNA | Bonnet monkey monomer I |
|---|---|---|---|
| I | 115 | 155 | (172) |
| II | 97 | 123 | 114 |

```
1                11               21               31
173              183              193              203
I   CCAAGGAAAT   ATCCTCCGAT       AACAAAGAGA       AAGAAGCTTT
II  AAAAGGAAAT   ATCCTCAGAT       GAAATCTGGA       AAGAAGCTTT

41               51               61               71
213              223              233              243
    CTGAGAAACT   TCTTTGTGTT       CTGTGAAATC       ATCTCACAGA
    CTGAGAAACT   GCTTAGTGTT       CTGTTAATTC       CTCTCGCAGA

81               91               101              111
253              263              273              283
    GTTACAGCTT   CCCCCTCAAG       AAGCCCTTCG       CTAAGACAGT
    GTTACATCTG   TATTTCGTGG       ATCTCTTTGC       TAGCCTTATT

121              131              141              151
293              302              312              322
    TCTTGTGGAA   TTGGCAAAGT       GATATTTGGA       AGCCCATAGA
    TC TGTGGAA   TCTGAGAACA       GATATTTCGG       ATCCCTTTGA

161              171
332              342
    GGGCTATGGT   GA
    AGACTATAGG   G
```

**Fig. 52.** Nucleotide sequence of the bonnet monkey 342 bp dimer (Rubin et al. 1980). There is a deletion at position 295 in monomer II. A palindrome (position 25–60) and a conservative sequence (G–A–A–A–T–A–T–C; positions 6–12 and 178–184) are observed in the sequence. One should also note the presence of a partial complement of the conservative sequence (positions 141–147, 263–268, 289–294, and 313–320)

### 6.5.2.2 Evolution

According to Donehower et al. (1980), primate stDNA evolution proceeded on the basis of unequal crossing-over, occurring both within and between the 170 bp repeats. A comparison of the nucleotide sequences of primate alphoid stDNA repeat units indicates the irregular character of divergence. Conservative regions alternating with the extensively diverged ones are encountered. Figure 51 shows relatively constant regions of primate DNA, alternating with the more "variable" regions. Donehower and coauthors reported eight crossing-over sites within the baboon *papio* 1 repeat, six such sites are assumed in the 171 bp repeat of humans. These sites are located at the border between the constant and the variable regions. The sequence

$$5'\ldots \text{AAGG}$$
$$\text{TTCC}\ldots 5'$$

or its invert

$$5'\ldots \text{TTCC}$$
$$\text{AAGG}\ldots 5'$$

is near these sites.

The authors assume that these short sequences could serve as recognition sites for specific nucleases probably participating in mammalian DNA recombination processes. At the same time, a comparison of the primary structures unambiguously demonstrates that human, bonnet monkey, and baboon stDNAs originated from the amplification of dimers consisting of related, but nonidentical sequences. The AGM α-stDNA is a result of 172 bp monomer amplification.

### 6.5.3 Human stDNA

#### 6.5.3.1 General Characteristics

A number of stDNAs have been revealed in the human genome (see Fig. 6). The buoyant density of the stDNAs in CsCl and their percentage are given in Table 17.

StDNA I is detected in human total DNA when the analytical ultracentrifuge cell is overloaded up to 25 µg (usually 2–5 µg of DNA are analyzed) (Corneo et al. 1967). StDNAs II and III were discovered using the $Ag^+ - Cs_2SO_4$ density gradient (Corneo et al. 1970b, 1971).

StDNA IV was isolated from total DNA by preliminary chromatography on the methylated albumin-kieselguhr column followed by ultracentrifugation in the $Ag^+ - Cs_2SO_4$ density gradient (Corneo et al. 1972).

Saunders et al. (1972) isolated satellite C with a density of $1.703 \, g/cm^3$. Later Chuang and Saunders (1974) discovered two more satellites, A and B. StDNAs A, B, and C, like the other stDNAs, were isolated in the $Ag^+ - Cs_2SO_4$ density gradient. It is noteworthy that though the presence of the A, B, and C components in the human genome was never disproved, they are rarely mentioned in later works devoted to human stDNAs. As a rule, discussions concern stDNAs I–IV.

**StDNA I.** Satellite I has been little investigated. It is digested with *Hinf*I and forms three fragments of 770, 850, and 950 bp. The 770 bp fragment contains recognition sites for four different restriction endonucleases, while the 850 and 950 bp fragments are resistant to the used restriction endonucleases (Frommer et al. 1982).

**StDNA II.** Satellite II is almost completely digested with *Hinf*I into small 10–80 bp fragments (Frommer et al. 1982). The size of the fragments does not indicate a defined regularity. It is shown that stDNA II contains *Hinf*I sites (G↓ANTC), a considerable part of which contains G or C in the center. Satellite II is also digested with *Taq*I with the formation of 35–190 bp fragments.

**StDNA III.** Satellite III is almost fully digested with *Hinf*I, the size of the formed fragments being 15–250 bp. It is shown that in contrast to satellite II, only A or T is found in the center of the *Hinf*I sites contained in satellite III.

**Table 17.** Human DNA components (composed from available literature data)

| DNA component | Buoyant density in CsCl, g/cm$^3$ | Percentage |
|---|---|---|
| Main component | 1.700 | — |
| stDNA I | 1.687 | 0.5 |
| stDNA II | 1.693 | 2.0 |
| stDNA III | 1.696 | 1.5 |
| stDNA IV | 1.700 | 2.0 |
| stDNA A | 1.710 | 0.5–1.0 |
| stDNA B | 1.726 | — |
| stDNA C | 1.703 | — |

Fragment
size

```
               5      10   15↓  20↓  25↓  30     35    40
45 bp    3' G--GGG--C|ACCTT|ACCTT|ACCT-|ACCTT|ACĞTT|ACCTT|ACC

               5      10   15     20     25     30     35    40     45
50 bp    3' GCCGGGCTC|ACGTC|ᴬCCTT|ACCTT|ACCTT|ACCTT|ACGTT|ACCTT|ACC
                                 C

               5      10     15     20     25     30     35    40     45
100 bp   3' G--GGG--C|ACGTT|ACCTT|ACCT-|ACCTT|ACCTT|ACCTT|ACCTT|ACCTT
               50     55     60     65     70     75     80     85    90
               GG-GGG---|ACCTT|ACCTT|ACCTT|ACCTT|ACCTT|A--TT|ACC
```

**Fig. 53.** Nucleotide sequence of the 45 bp, 50 bp, and 100 bp *Hinf*I fragments of human stDNA III (Frommer et al. 1982). The 5 bp TTCCA monomer is marked, the positions diverging from the consensus sequence are *underlined*. The 100 bp fragment is represented as two tandem repeats of the 49 bp unit

The absence of a *Taq*I site in satellite III is one of the indications distinguishing it from satellite II.

The primary structure of three *Hinf*I 45 bp, 50 bp, and 100 bp fragments from satellite III has been determined (Fig. 53). All three sequences contain the tandemly repeating pentanucleotide TTCCA ending with a nine bp GC-enriched sequence. The 100 bp sequence is a dimer, having lost the *Hinf*I site as result of a one base deletion.

### 6.5.3.2 *Hae*III and *Eco*RI Analysis of stDNAs II and III

Several publications have presented data indicating that the human satellites II and III are digested with *Hae*III or *Eco*RI and yield a "ladder" structure. According to Frommer et al. (1982) these fractions represent the minor component in stDNA preparations. Moreover, according to Corneo et al. (1982), satellite III is practically not digested with *Eco*RI and *Hae*III if the stDNA is thoroughly purified controlling the purification by reassociation kinetics and strand separation in the alkaline CsCl gradient. The source of these fractions can be the so-called homogeneous main band (HMB) which can be isolated by gradient ultracentrifugation. StDNA II resistance to *Eco*RI and *Hae*III has been observed also by Manuelidis (1978).

This circumstance compels a cautious approach to the interpretation of earlier obtained results on the localization of satellites II–IV in chromosomes and their structural organization.

Mitchell et al. (1979) observed the formation of a 170 bp monomer and its multiple oligomers at digestion of satellites II and III with *Hae*III restriction endonuclease. A 3.4 kb fragment was also detected. This fragment is absent both in stDNA and in female total DNA (Bostock et al. 1978). The "male" fragment is not detected in satellites I and IV. It is located in the long arm of the Y-chromosome.

*Eco*RI digestion of satellites II and III yields a number of fragments differing in their relative content (Mitchell et al. 1979). The length of the short fragments formed at digestion with *Eco*RI is 340 and 680 bp. The long 1.77 and 3.4 kb fragments were cloned (Cooke and Hindley 1979). It was shown that these fragments no not hybridize

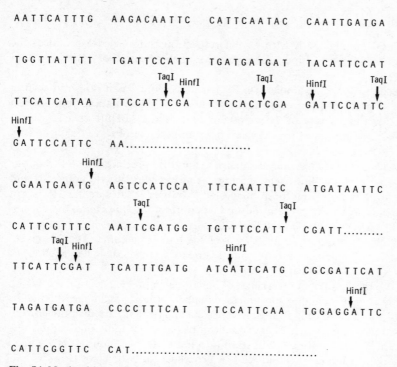

**Fig. 54.** Nucleotide sequence of the 1.77 kb fragment of human stDNA III (Cooke and Hindley 1979)

with each other and that the 3.4 kb *Eco*RI fragment hybridizes exclusively with male DNA.

The 1.77 kb fragment was cleaved with *Alu*I and the primary structure of three segments was determined, one from each end of the fragment and one from the center. The overall length of the determined sequence was 300 bp (Fig. 54). A number of repeating elements are encountered in the sequence, $ATTC_G^CATTC$ being the most frequent. It is observed four times with the longer related sequences which occur only once: ATTCCATTCGATTC and ATTCCATTCGATTCATTC.

Sequence ATTCGATTC contains overlapping *Hinf*I and *Taq*I sites as a result of which the 1.77 kb fragment at digestion with these enzymes produces many short sequences.

The presence of a conservative segment, alternating with the less conservative ones of different length, does not permit classifying the 1.77 kb sequnce as a very simple or complex one as it contains elements belonging to both categories.

The 3.4 kb *Eco*RI fragment has been much less studied. It is apparently internally repetitive and heterogeneous. According to Cook and Hindley, at least three or probably more not linked covalently subsatellite fractions are contained in satellite III. These components differ in sequence and in their localization in chromosomes.

### 6.5.3.3 Alphoid stDNA

Digestion of total human DNA with *Eco*RI restriction endonuclease produces the 340 bp fragment and its multiple oligomers (Manuelidis and Wu 1978; Wu and Manuelidis 1980).

This sequence is not an example of a "classical satellite" as it does not form a separate band at gradient ultracentrifugation. However, its structural organization – the tandemly arranged repeats – is close to that of the satellite (Darling et al. 1982). A similar fraction from the African green monkey is observed as a density stDNA.

It should be noted that digestion of satellites II and III produces *Eco*RI fractions of analogous length. A low, but true homology, is observed between satellites II and III and alphoid stDNA. It has been suggested, however, that satellite III is not isolated in the pure form and contains admixtures both of satellite II and the considered alphoid DNA (Wu and Manuelidis 1980).

Alphoid stDNA composes about 0.75% of the human genome. The 340 bp sequence contains two tandem repeats 169 and 171 bp long (Fig. 55). 46 bases (27%) of these sequences are not identical.

In contrast to the significant divergence of individual subrepeats within the 340 bp dimer, the dimer itself in higher multimers remained almost unchanged. Thus, as compared with the dimers, only four base replacements ($< 1\%$) are distinguished in the 680 bp tetramer. It is evident that sequence divergence of individual subrepeats took place in the dimer before amplification into long stDNA chains. It is assumed that the formation of human alphoid stDNA proceeded in two stages (Fig. 56).

```
    1              11             21             31
    172            182            192            202
I  AATTCTCAGT     AACTTCCTTG     TGTTGTGTGT     ATTCAACTCA
II GATTCTCAGA     AACTCCTTTG     TGATGTGTGC     GTTCAACTCA

    41             51             61             71
    212            222            232            242
   CAGAGTTGAA     CGATCCTTTA     CACAGAGCAG     ACTTGAAACA
   CAGAGTTTAA     CCTTTCTTTT     CATAGAGCAG     TTAGGAAACA

    81             91             101            111
    252            262            272            282
   CTCTTTTTGT     GGAATTTGCA     AGTGGAGATT     TCAGCCGCTT
   CTCTGTTTGT     AAAGTCTGCA     AGTGGATATT     CAGACCTCTT

    121            131            141            151
    292            302            312            321
   TGAGGTCAAT     GGTAGAATAG     GAAATATCTT     CCTATAGAAA
   TGAGGCCTTC     GTTGGAAACG     GGATT TCTT     CATATTATG

    161            171
    330            340
   CTAGACAGAA     T
   CTAGACAGAA     G
```

**Fig. 55.** Nucleotide sequence of the 340 bp segment formed at cleavage of total human DNA with restriction endonuclease *Eco*RI (Wu and Manuelidis 1980). *Numerals I and II* indicate the 171 and 169 subrepeats

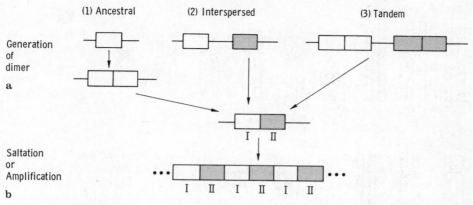

**Fig. 56.** Hypothetical scheme of human alphoid stDNA formation (Wu and Manuelidis 1980). At least two steps are assumed: the formation of a dimer (**a**) and saltatory replication or amplification of a dimer in stDNA (**b**). *Numerals* I and II denote monomers. It is presumed that there can be three possible variants of the initial dimer. (1) The 170 bp initial sequence is duplicated producing a dimer and sequence divergence occurs; (2) two diverged sequences are rejoined by unequal crossing-over or by any other recombination event; (3) tandem variants within the genome recombine, forming a dimer with related, but not identical sequences

The structure of human alphoid DNA fragments is to a considerable degree similar to that of the 172 bp *Hind*III fragment of the AGM (Rosenberg et al. 1978). 65% of the human and AGM 171 bp sequences are identical. The difference is in the insert of one base into the monkey genome. This extra residue (guanine) is located between nucleotides 125 and 126. A cluster of base replacements in the center of the molecule is also observed. The 340 bp repeat of human alphoid stDNA contains many terminating codons when the sequence is read in both directions (7–8 in each direction). One usual initiating codon (ATG) is encountered as well as an unusual initiator (GTG) (3–5 in each direction). The presence of several symmetry and dyad symmetry regions have also been noted.

Analysis of the primary structure of the human alphoid stDNA repeat does not provide any convincing data to corroborate their formation from short sequences.

Human alphoid stDNA from different organs show a different level of methylation. The $m^5C$ content ratio from brain, placenta, and sperm is $2.0 : 1.2 : 1.0$ (Gama-Sosa et al. 1983).

### 6.5.3.4 Chromosomal Location

20% of the human genome is heterochromatic. The bulk of heterochromatin represents centromeric heterochromatin. The Y-chromosome is an exception, revealing a large heterochromatic segment in the longer arm (Miklos and John 1979). Since the amount of human stDNAs is much less than that of heterochromatin, it is evident that most of the heterochromatic DNA is nonsatellite.

It has been shown that about 23% of human DNA reassociates at $C_0t = 1$. But it has not yet been determined whether all this DNA is localized in the hetero-

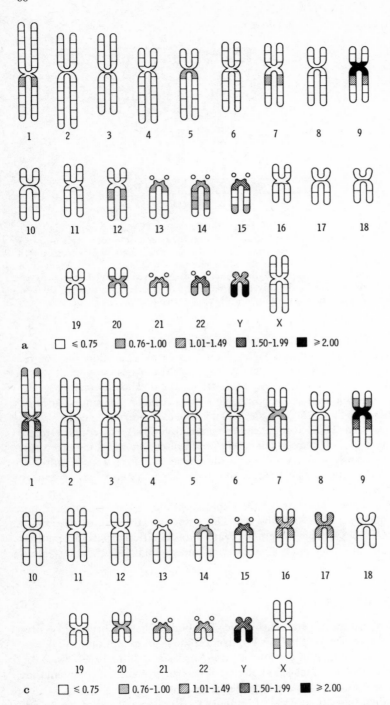

**Fig. 57.** Satellite cRNA hybridization with human chromosomes (Gosden et al. 1975). **a** stDNA I; **b** stDNA II; **c** stDNA III; **d** stDNA IV. The *symbols* indicate the number of grains per segment

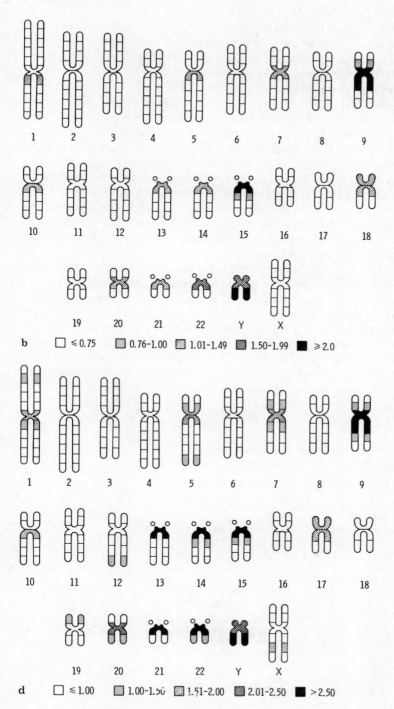

b  □ ≤ 0.75  ▨ 0.76–1.00  ▨ 1.01–1.49  ▨ 1.50–1.99  ■ ≥ 2.0

d  □ ≤ 1.00  ▨ 1.00–1.50  ▨ 1.51–2.00  ▨ 2.01–2.50  ■ > 2.50

chromatin. Marx et al. (1976) isolated five fractions containing repeating sequences from total human DNA by reassociation and subsequent fractionation of the reassociated molecules on hydroxyapatite. The buoyant densities of three light components (1.687, 1.696, 1.700) and satellites I–III coincide.

Localization of satellites I–IV was determined with $^3$H-cRNA transcribed from the stDNAs (Gosden et al. 1975). All four stDNAs hybridized mainly with chromosomes 9 and Y, but the label was detected also in the centromeric regions of chromosomes 1, 5, 7, 10, 12, 13, 14, 17, 20 21, and 22. However, only cRNA, transcribed from stDNA II, hybridized with the centromeric region of chromosome 16. It is not in every chromosome that stDNAs are found in the centromeric heterochromatin (Fig. 57). Quantitative data on the in situ hybridization of stDNAs I–IV are given in Table 18.

The fact that cRNAs from each stDNA hybridize with the heterochromatin of many chromosomes is explained by the authors as follows: (1) the centromeric regions of these chromosomes contain adjoining blocks of different stDNAs, or (2) heterologous hybrids are formed between the stDNAs, and, consequently, the cRNA of every satellite hybridizes with different sites.

Mitchell et al. (1979) analyzed the sequence homology of human stDNA I–IV nucleotide sequences. It was shown that satellites III and IV in some properties are indistinguishable from each other (buoyant density, reassociation, and others). 40% of the satellite III sequence is homologous to 10% of satellites I and II. As concerns satellites I and II, no homology is observed between them. An explanation of the fact that cRNA, synthesized on individual human stDNAs, hybridize to the same regions of chromosomes is provided by the authors in the light of these data. This can be due to the homology between the stDNAs, but this does not automatically repudiate the first suggestion by Gosden and coauthors.

**Table 18.** Satellite cRNA hybridization with human chromosomes (% of autoradiographic grains per segment) (Gosden et al. 1975)

| Chromosome | I | II | III | IV |
|---|---|---|---|---|
| 1 | 1.06 | 13.70 | 1.76 | 3.75 |
| 5 | 1.06 | 0.00 | 1.40 | 1.94 |
| 7 | 1.32 | 0.36 | 0.35 | 2.01 |
| 9 | 42.74 | 27.04 | 39.63 | 33.94 |
| 10 | 0.00 | 0.00 | 0.35 | 0.80 |
| 12 | 1.06 | 0.00 | 0.00 | 0.13 |
| 13 | 2.11 | 0.00 | 2.46 | 4.96 |
| 14 | 5.28 | 1.59 | 4.56 | 6.16 |
| 15 | 7.91 | 6.85 | 14.04 | 10.52 |
| 16 | 0.00 | 7.93 | 0.00 | 0.00 |
| 17 | 0.00 | 4.69 | 0.70 | 3.62 |
| 19 | 0.00 | 0.00 | 0.00 | 0.94 |
| 20 | 1.58 | 0.72 | 3.16 | 3.22 |
| 21 | 5.28 | 2.88 | 4.56 | 5.36 |
| 22 | 8.44 | 3.96 | 7.37 | 7.97 |
| X | 0.00 | 1.44 | 0.00 | 0.13 |
| Y | 22.16 | 28.84 | 19.66 | 14.55 |

The simultaneos hybridization of total stDNAs with several chromosomes can indeed result from localization of individual fractions in defined chromosome pairs. It has been shown, for example, that total stDNA III contains a set of sequences, some of which are predominantly localized only in defined chromosome pairs (Gosden et al. 1981). Thus, a 1.77 kb cloned *Eco*RI-fragment of human stDNA III is hybridized mainly with chromosome 1. On the whole, satellite III contains at least three sequences: the "Y-chromosome specific" and the "1-chromosome specific", in addition to the bulk stDNA III.

Beauchamp et al. (1979) developed a new and sensitive technique of determining the chromosomal localization of human stDNA. A DNA isolated from different mouse-human somatic cell hybrids containing one or several human chromosomes is applied to membrane filters and hybridizes with radioactive restriction fragments of individual stDNAs. This method can determine chromosomal localization not only of a defined stDNA, but of its different sequences as well. The sensitivity of this method is very high and permits detection of very small sequences occupying as little as $2 \cdot 10^{-5}\%$ of a genome.

Using this procedure, Beauchamp and coauthors detected sequences homologous to satellite III in chromosomes 1, 7, 11, 15, 22, and X. Distribution of restriction endonuclease sites within a stDNA III is different on different chromosomes. Furthermore, the high sensitivity of the method revealed regions in the X-chromosome complementary to satellites I, II, and III. No stDNAs have been detected previously in the X-chromosome.

## 6.6 Plant stDNA

StDNAs in higher plants were discovered in 1967. Several groups demonstrated independently the presence of a minor component in the french bean (*Phaseolus vulgaris*) DNA (Beridze et al. 1967; Matsuda and Siegel 1967; Meyer and Lippincott 1967). This component was detected in nuclei precipitated through 2.2 M sucrose and free from admixtures of cytoplasmic organelles. The shape of stDNA melting curves showed that it consists of two components (Beridze et al. 1967). These results were confirmed later by Wolstenholme and Gross (1968). From that time and up to 1972 no data were reported in the literature on stDNA detection in other high plant species.

The discovery of stDNAs in plants evoked a cardinal reconsideration of conceptions on the mechanism of DNA component distribution in plant cell organelles. The main concern being that in most cases cpDNA ($\varrho = 1.697 \text{ g/cm}^3$) coincides in density with the plant main nDNA component and is not detected as a density component in the gradient. For many years mtDNA ($\varrho = 1.706 \text{ g/cm}^3$) was erroneously identified as a cpDNA. The detection of a minor DNA component in plant nuclear fractions and the coincidence of cpDNA and nDNA density led to greater caution in considering problems of intracellular localization of DNA minor fractions. The story of identification of plant DNA minor components is fascinatingly described in a review paper by Kirk (1971). In the section entitled "Satellites Disappear" Kirk wrote: "In 1967 there began to appear a series of papers, which changed the whole picture. The first of them, by Beridze et al., reported studies by the CsCl technique on DNA from nuclei and chloroplasts of *Phaseolus vulgaris*, the French bean" (p. 269).

### 6.6.1 Interspecies Distinction

It was first shown with the *Phaseolus* genus as an example that closely related higher plant species within one genus can differ in the quantity of density stDNAs (Beridze 1972). A distribution study in the CsCl density gradient was done of nDNAs of seven species of the genus *Phaseolus* from two different zones of primary cultivation of cultured bean species, Central America and South-East Asia. The diploid number of chromosomes of all the bean species was $2n = 22$ (Zhukovsky 1971). The stDNA content in all the American species is about 30% of the total amount of nDNA. In two Asiatic species the stDNA content, in my calculation, was 10–15%, while *P.aureus* did not reveal any stDNA. The observed differences were mainly confirmed later by two groups (Ingle et al. 1973; Seshadri and Ranjekar 1979).

Marked differences in the quantitative content of stDNA were found in species of the *Brassica* family with different chromosome numbers (Beridze 1975). In *B.oleracea* ($2n = 18$) the density stDNA ($\varrho = 1.704\,g/cm^3$) is absent, in *B.campestris* ($2n = 20$) its content is 24%, and in *B.nigra* ($2n = 16$) it composes 37%. These species with different contents of stDNA form allopolyploids with a quantitatively intermediate stDNA amount.

Data on the stDNA content in citric plants are also of interest. The stDNA of citric plants attract attention both by their high percentage content and the highest buoyant density among the analyzed plants, $1.711–1.712\,g/cm^3$ (Ingle et al. 1973; Bragvadze and Beridze 1976, 1983). The citric plants studied in our work involve four genera (14 species). The stDNA content varied within the limits of 10–30%.

Interesting results were obtained from a study of nDNA from *Poncirus trifoliata*, used as rootstock in citric plant breeding. This species hybridizes with other citric plants, but the obtained hybrids are sterile. In *P.trifoliata* we detected two stDNAs

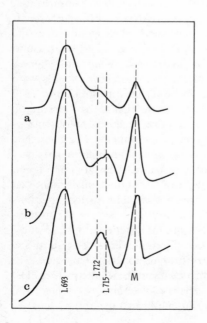

**Fig. 58.** Distribution of nDNAs of different species from the *Citrinae* subtribe in the CsCl density gradient (Bragvadze and Beridze 1983). **a** *Citrus limon*; **b** *Poncirus trifoliata*; **c** *Citrus x Poncirus* hybrid

with densities of 1.712 and 1.715 g/cm³. The *Citrus X Poncirus* hybrid plants are also characterized by two stDNAs (Fig. 58).

Ingle et al. (1973) examined many higher plant species belonging to different taxonomic groups. Their main conclusion was that density stDNAs are characteristic of dicotyledonous plants. Of the 60 dicotyledonous plant species analyzed, density stDNAs were observed in 27, while in all of the 11 monocotyledonous plants studied they were absent. In another study the same authors reported the absence of density stDNAs in 15 more monocotyledonous plants (Ingle et al. 1975). In their opinion, the presence of density stDNAs is not an indication of a greater primitiveness or more advancement of dicotyledonous plant families.

StDNAs were most frequently encountered in two families, the *Cucurbitaceae* and *Rutaceae*. However, these families also contained species without stDNAs. The authors draw the conclusion that the presence of density stDNAs is correlated neither with the accepted schemes of evolution of flowering plants nor with their taxonomic classification. It was later demonstrated that monocotyledonous plants also contain density stDNAs (Capesius et al. 1975; Capesius 1976).

It should be mentioned that only in two cases of 27, the stDNA density is less than that of the DNA main component (*Lobularia maritima*, 1.688 g/cm³, *Linum usitatissimum*, 1.689 g/cm³), i.e., stDNAs exhibit a tendency to a high GC content.

## 6.6.2 Nucleotide Sequence

Plant stDNAs were revealed at the same time as animal stDNAs, but their further studies were less intensive. At present stDNAs have been characterized for several higher plant species (Bendich and Anderson 1974; Chilton 1975; Sinclair et al. 1975; Beridze and Bragvadze 1976; Capesius 1976, 1979; Timmis and Ingle 1977; Deumling and Nagl 1978; Hemleben et al. 1977; Grisvard and Tuffet-Anghileri 1980; Fodor and Beridze 1980a; Shmookler Reis et al. 1981; Hemleben et al. 1982; Bragvadze 1983). Hidden stDNAs from monocotyledonous and dicotyledonous plants were also isolated and examined (Deumling et al. 1976; Ranjekar et al. 1978; Wall and Bryant 1981).

The nucleotide sequence was determined only for several plant stDNAs. Dennis et al. (1980) determined the nucleotide sequence of the repeating element of wheat and barley hidden stDNA ($\varrho = 1.699$ g/cm³). The stDNA is fully digested with *Mbo*II restriction endonuclease producing a series of discrete fragments 3, 6, 9, 12, 15, ... bp in length. The size of the longest fragment did not exceed 50 bp. More than 90% of stDNA was digested, producing fragments shorter than 21 bp. Analysis of cRNA synthesized on stDNA confirmed data on the simplicity of wheat and barley stDNA structural organization and permitted to establish the structural formula as $(GAA)_m(GAG)_n$.

Deumling (1981) determined the nucleotide sequence of stDNA repeat units in the *Scilla sibirica* monocotyledonous plant. StDNA digestion with *Hae*III restriction endonuclease generated a series of fragments ranging from 33 to several hundred bp in length. The nucleotide sequence was determined for four main fragments (33, 48, 57, and 67 bp in length). The main element of the *S.sibirica* stDNA is a 33 bp fragment representing a palindrome (Fig. 59). This sequence in stDNA is repeated tandemly

```
          M      *  *M  M      M  *M
   5'- CCCATGCACC GAACCGCCCG CGGCTCGTCC GTGGG
```

**Fig. 59.** Nucleotide sequence of *Scilla sibirica* stDNA 33 bp fragment (**a**) and one of the chains of the same sequence represented as a hairpin and indicating that the main repeat is a palindrome (**b**) (Deumling 1980). Fully methylated cytosine residues are marked with the letter M, the partially methylated ones with an *asterisk*

```
             M     M    M          *      *
 a    GGGTACGTGG CTTGGCGGGC GCCGAGCAGG CACCC -5'


                      C                C
   3'- G G G T G C     T G   C T     G G C G
                                               C
         I I I I   I    I I   I I   I I I I C
   5'- C C C A T G   A C   G A   C C G C   C
 b                 C       C     A
```

```
G G C C A C A C A A C C C C C A T T T

T G T C G A A A A T A G C C A T G A A C

G A C C A T T T T C A A T A A T A C C G

A A G G C T A A C A C C T A C G G A T T

T T T G A C C A A G A A A T G G T C T C

C A C C A G A A A T C C A A G A A T G T
Sau3A
[G A T C] T A T G G C A A G G A A A C A T

A T G T G G G G T G A G G T G T A T G A
         TaqI      Sau3A
G C G T C T G G [T C G A] [G A T C] A A T

G G C C
```

**Fig. 60.** Nucleotide sequence of the corn stDNA cloned monomer (Peacock et al. 1981). *Taq*I and *Sau*3A restriction sites are *boxed*

with some base substitutions and an insertion of the tetranucleotide

$$5'\dots GTTC$$
$$CAAG\dots 5'.$$

Peacock et al. (1981) have established also the nucleotide sequence of corn hidden stDNA which is localized in knob heterochromatin. The repeating sequence of this stDNA, consisting of 180 bp, does not contain noticeable internal repeats (Fig. 60).

The primary structure of one of the lemon stDNA *Bsp*I fragments has been determined (Beridze and Shengeliya, unpublished results). Lemon stDNA is characterized by a high GC content (about 65%) with the resulting high buoyant density value of 1.712 g/cm$^3$ in the neutral CsCl density gradient (Beridze 1980c). Restriction endonuclease digests lemon stDNA into 16 fragments with a length of 10–200 bp. The primary structure of one of the *Bsp*I-fragments, 90 bp in length, is shown in Fig. 61.

### 6.6.3  5-methylcytosine Content

In a number of cases studies of plant stDNA properties showed disagreement in the GC content values derived from the $T_m$ and buoyant density in CsCl, e.g., the GC content of lemon stDNA is 53.1 and 65.4%, respectively (Beridze 1980b). This

```
1                11               21
CCGCCTTGAA       TCGTAATTCC       ATCGAGCGGC

31               41               51
GGGTAGAATC       CTTTGCAGAC       GACTTAAATA

61               71               81
CGCGACGGGG       TATTGTAAGT       GGCAGAGTGG
```

**Fig. 61.** Nucleotide sequence of the lemon stDNA 90 bp cloned *Bsp*I fragment (Beridze and Shengeliya, unpublished)

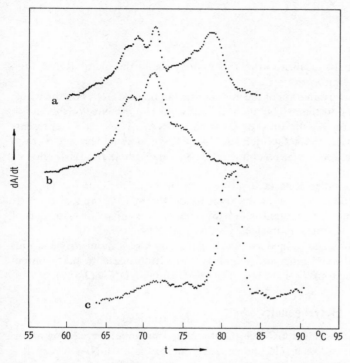

**Fig. 62.** Differential melting curves of plant stDNAs (Beridze 1980b). **a** *Phaseolus coccineus*; **b** *Brassica nigra*; **c** *Citrus limon* (0.1 × SSC)

discrepancy may be an indication of the large content of the minor base, 5-methylcytosine, in the satellite component. Kemp and Sutton (1976) proposed a formula for estimating the content of this minor base proceeding from the $T_m$ and buoyant density in CsCl. It is assumed that $m^5C$ decreases the density, but does not affect $T_m$. Calculations show that the stDNA of lemon and bean must contain a significant amount of $m^5C$, whereas its content in the stDNA of *Brassica nigra* is evidently less than in the main component (Beridze 1980b).

StDNAs of *Phaseolus* and *Citrus* as well as all the "heavy" satellites of the studied plants, with the exception of *B.nigra*, contain a high temperature melting component (Bendich and Anderson 1974; Chilton 1975; Timmis and Ingle 1977; Beridze and Bragvadze 1976; Beridze 1980b) (Fig. 62). The low $m^5C$ value in the *B.nigra* stDNA

**Table 19.** Base content of lemon DNA fractions (number of determinations 3–5)

| Fraction | Mol % of total bases | | | | | |
|---|---|---|---|---|---|---|
| | T | G | A | C | m⁵C | GC |
| Total DNA | 27.5 | 22.6 | 27.3 | 20.5 | 2.2 | 45.3 |
| Satellite DNA | 16.1 | 34.7 | 15.4 | 25.0 | 9.8 | 69.5 |
| Main component | 30.1 | 20.2 | 29.5 | 16.8 | 2.3 | 39.3 |

indicates the predominant localization of this minor base in the plant stDNA high temperature melting component.

A direct chemical determination of the $m^5C$ content in plant stDNAs on the whole confirmed indirect data on the $m^5C$ content in stDNAs. The presence of this minor base proved to be characteristic only of GC-rich stDNAs. Thus, $m^5C$ represents about 25% of all the bases in the GC-rich stDNA of *Scilla sibirica*. The $m^5C/C$ ratio for the total stDNA is about 1.5, reaching 2.2–2.8 in individual *Hae*III-fragments (Deumling 1981).

According to Shmookler Reis et al. (1981), the $m^5C$ content in the GC-rich stDNA of melon composes 11.8% of the total bases, the $m^5C/C$ ratio being 0.76.

In lemon stDNA the $m^5C$ content is fourfold higher in comparison with that of the main component (Mazin, unpublished results) (Table 19).

The $m^5C$ content was determined in an AT-rich stDNA of a monocotyledonous plant *Cymbidium pumilum* (Wagner and Capesius 1981). In this case the $m^5C$ content was lower (1.4% of the total bases) than in the total cellular DNA (3.5%).

### 6.6.4 Intermolecular Heterogeneity

Up to the present all the studied GC-rich plant stDNAs are characterized by multi-component melting curves. As a rule, the reassociation of these stDNAs reveals fast and slowly reassociating fractions.

The heterogeneity observed by methods of thermal denaturation and reassociation kinetics may reflect both the intra- and intermolecular heterogeneity. Equilibrium centrifugation in the actinomycin D-CsCl and $Hg^{2+} - Cs_2SO_4$ density gradients is reliable evidence for the existence of intermolecular heterogeneity in stDNAs. These methods reveal the presence of three to four components in all of the stDNAs studied (Beridze 1980b).

It is known that rDNA is localized in the density zone of heavy stDNAs (Ingle et al. 1975). Calculations show that about 10–20% individual stDNAs can represent rDNA. In some cases direct experiments have identified rDNAs among plant stDNA components (Hemleben et al. 1977; Fodor and Beridze 1980b; Bendich and Ward 1980).

Another reason for the intermolecular heterogeneity observed in plant stDNA preparations is the admixture of mtDNA which is close in buoyant density to stDNA and rDNA. The stDNA of melon was isolated earlier into two components. One consisted of simple sequences (kinetic complexity $- 0.3 \cdot 10^6$ dalton) and was repeated $4 \cdot 10^6$ times, while the other was 6000 times more complex and occurred only $10^2$ times in the genome (Bendich and Anderson 1974; Sinclair et al. 1975; Bendich and Taylor 1977). It was later conclusively demonstrated, however, that the complex component is a mtDNA (Grisvard and Tuffet-Anghileri 1980; Bendich and Ward 1980). Such a fact is another reminder that intracellular localization of DNA minor fractions must be stringently defined before their detailed study.

# 7 Structure of stDNA-Containing Chromatin

It is known that staining of eukaryotic metaphase chromosomes with basic dyes reveals an irregular and strictly specific binding of the stains along the chromosome. This is explained by the character of irregular DNA packing along the chromosome length, i.e., at defined sites of the chromosomes the amount of DNA per unit volume is greater or less than in the other sites. The highest DNA specific concentration is assumed to correspond to the sites of the constitutive heterochromatin.

The condensed state of the constitutive heterochromatin is apparently somehow connected with the presence of stDNAs. The principal and, probably, the only function of a stDNA may be its capability to form the constitutive heterochromatin.

How is constitutive heterochromatin organized? Are there any specific proteins present in it? What is the structure of stDNA-containing nucleosomes? These are the questions which still require a final solution.

Some reports have appeared in recent years which could shed light on the structural organization of heterochromatic regions of chromosomes containing stDNA.

## 7.1 Isolation of Satellite Chromatin

Achievements in studies of stDNA-containing chromatin (i.e., satellite chromatin) resulted from the use of restriction endonucleases for their isolation. In earlier heterochromatin isolation studies, the ultrasonic technique was used to disrupt nuclei. After disruption the heterochromatin was pelleted by centrifugation, while the euchromatin fragments remained in the supernatant (Frenster et al. 1963). Such a procedure is inadequate, as a drastic disruption of the nuclei structure does not exclude the possibility of a change in the heterochromatin native structure, loss or modification of specific proteins, etc. The use of restriction endonucleases to isolate satellite chromatin does not have these drawbacks. Due to the homogeneity of stDNA structure, it is possible in some cases to select a restriction endonuclease that digests nonsatellite DNA, leaving the stDNA intact, or vice versa.

Naturally, preference should be given to the variant where the satellite DNA is undigested and the bulk chromatin is fragmented. In this case there is a much lesser chance to lose the specific heterochromatin proteins if such exist.

At present the most improved procedure is the two-step method of satellite chromatin isolation applied in mouse cell studies (Zhang and Hörz 1982). The method is based on the preliminary solubilization of chromatin with micrococcal nuclease and subsequent treatment of chromatin with restriction endonuclease *Bsp*I which digests nonsatellite DNA, but leaves stDNA intact. Due to its rapid sedimentation, the separation of satellite chromatin from the digested remaining chromatin is done by centrifugation in the sucrose density gradient. As a result, about 70 % of the DNA is accounted for the satellite component.

It should be noted here that satellite chromatin, as a rule, contains some part of the DNA which coincides in density with the main component. Some differences were detected between this fraction and total bulk DNA in the case of mouse chromatin (Lica and Hamkalo 1983). This question has been little studied, and thus, nonsatellite DNA of satellite chromatin are objects for future studies, without which it is impossible to create a full picture of the stDNA-containing chromatin structural organization.

## 7.2 Reasons for the Compactness of Satellite Chromatin

As mentioned above, satellite chromatin is in a more compact dense state than the bulk chromatin, both in the interphase nuclei and in the metaphase. The density of DNA in heterochromatin is 1.5 times greater than in euchromatin (Bedbrook et al. 1980). Physicochemical analysis corroborates this conclusion. Thus, the melting point of stDNA is the same as that of the main component at thermal chromatography of mouse liver chromatin on hydroxyapatite, but pure stDNA in solution melts at lower temperature than the main component due to the lower GC content (Pashev and Markov 1978).

As already indicated, polynucleosomes containing stDNA sequences are digested slower by micrococcal nuclease than the rest of the chromatin polynucleosomes which also must be a consequence of the condensed state of chromatin (Zhang and Hörz 1982).

What determines the condensed state of stDNA-containing chromatin?

The noncovalent interaction of the constituents, which are very sensitive to the ionic medium, can play an important role in determining the compact state of heterochromatin. Thus, it has been shown that mouse satellite chromatin binds calcium ions more intensively than bulk chromatin does. A high concentration of univalent ions, on the contrary, enhances solubilization of satellite chromatin (Zhang and Hörz 1982). The condensed state of satellite chromatin can also be determined by specific proteins, similar to those detected in *D.melanogaster* (Blumenfeld et al. 1978; Hsieh and Brutlag 1979b, Levinger and Varshavsky 1982). Finally, the reason for the more compact packing of satellite chromatin, in contrast to the bulk chromatin, may be the specific phasing of nucleosomes along the stDNA chain. However, in any case, the key to the puzzle must be the chemical structure of DNA contained in the constitutive heterochromatin. The main task here, and perhaps of the whole problem of satellite DNAs, should be the elucidation of the structural features determining the compact state of constitutive heterochromatin, whether by mediation of specific proteins or by other factors.

## 7.3 Chromatin of the African Green Monkey

Musich et al. (1977a) isolated stDNA-containing nucleosoms during treatment of AGM nuclei with the *Eco*RI restriction endonuclease. The 172 bp length of DNA in the nucleosome coincides with the size of the DNA in the stDNA repeat units. At first the linker in the chromatin was assumed to be around the *Eco*RI site (position 30, Fig. 49), but later studies with the use of micrococcal nuclease identified the linker around position 126. It was shown that digestion of satellite chromatin by micrococcal nuclease takes place at position 126, whereas pure stDNA is digested in many other places (Musich et al. 1982).

The main difference between stDNA-containing nucleosomes and the bulk of nucleosomes is in the following: nonhistone proteins compose 65–70% of the proteins in the first type and 40% in the bulk nucleosomes. The amount of H1 histone is 13.7% in the satellite-free nucleosomes. The H1 histone is completely absent in the α-nucleosomes and instead of it five minor proteins of a similar molecular mass (24,000–43,000 daltons) were detected in the α-nucleosomes. Their amount was 12%. These proteins are absent in nonsatellite nucleosomes. The resistance to extraction (0.6–2.0 M NaCl) provides evidence that the low molecular weight nonhistone proteins are firmly bound with nucleosome components. It is worth mentioning, however, that other authors explain the absence of the H1 histone by its small size and consequent loss during isolation of satellite chromatin (Omori et al. 1980; Mathew et al. 1981).

It has also been suggested that a redistribution or degradation of the proteins could have occurred in these experiments, affecting the composition of the African green monkey satellite chromatin (Zhang and Hörz 1982).

Musich et al. (1978) and Brown et al. (1979) reported that the size of long stDNA repeats from different animals coincide, and are identical to the DNA length in the nucleosome of constitutive heterochromatin. It turned out that restriction analysis of stDNAs from a number of animals can reveal repeats similar to those of monkey α-stDNA (172 bp), even though they are observed as minor fractions in the cleavage products (Fig. 63). The authors assume that the occurrence of such repeats in the stDNA of different species cannot be a random one and that it reflects the general mechanism serving for the formation and evolution of the long-range periodicity. Nucleosomal organization could be the principal element in the determination of the stDNA long-range perodicity in the process of evolution.

In studies of the same African green monkey α-stDNA, Fittler and Zachau (1979), however, came to the conclusion that there is no simple phase relationship between nucleosomal and satellite repetition. The authors demonstrated that treatment of nuclei with micrococcal nuclease produces both 188 and 172 bp DNA fragments. The explanation could be that a part of the nucleosomes in monkey nuclei are organized in a phase correspondence with the repeating stDNA unit (phasing implies coincidence of the stDNA repeating element length with that of the nucleosomal DNA of satellite chromatin). However, it turned out that free stDNA also forms 172 bp fragments and multiple oligomers at incomplete cleavage with micrococcal nuclease.

A qualitative analysis of pure stDNA digestion products has shown that 50% of all digestions occur at positions 123 and 132, 5% at positions 79, and 1–3% at 20 other places (Hörz et al. 1983). They assume that it is not nucleosomal phasing, but the specifity of micrococcal nuclease to defined sequences that is responsible for the results of Musich and co-workers.

In experiments carried out by Fittler and Zachau (1979) the incubation of nuclei with DNase II yields repeats 188 bp long. In the stDNA register no bands were

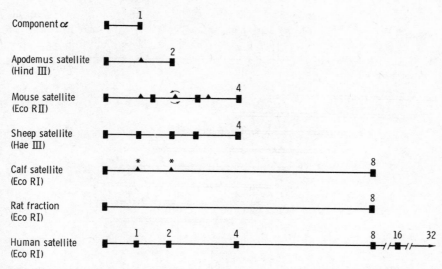

**Fig. 63.** Distribution of major and minor restriction sites in the stDNA of different animals (Musich et al. 1978). The repeating unit of the AGM α-stDNA (172 bp) is used as a standard. The major restriction sites are denoted by *squares*; the sites occurring less frequently and producing minor bands during gel electrophoresis are shown as *triangles*

observed. On the basis of these studies the authors conclude that the size of DNA in nucleosomes does not depend on the length of the stDNA repeats. According to Singer (1979), the DNA length in nucleosomes of satellite chromatin has a close value of 185 bp and does not differ from the bulk chromatin.

Zhang et al. (1983) working with AGM satellite chromatin have developed an original approach for determining the arrangement of nucleosomal frames along the satellite DNA chain. The procedure is as follows: core particles are extracted from the nuclei by subsequent treatment with micrococcal nuclease, exonuclease III, and nuclease S1. Satellite sequences are isolated from the core particles by reassociation of denatured DNA at low $C_0t$ values and chromatography on hydroxyapatite. Satellite core DNAs are then mapped relative to known restriction sites along the stDNA chain permitting to judge the positions of nucleosome frames in satellite chromatin. The quantitative correlation of individual frames in the chromatin is also determined.

An analysis of AGM satellite chromatin core particles has shown that nucleosomes occupy eight strictly defined positions in the α-stDNA repeat (Zhang et al. 1983). One of the frames composes 35% and all the others 10% each (Fig. 64). Micrococcal nuclease digestion reveals only one F-frame, as this nuclease in conditions of limited digestion and low temperature attacks only the two most sensitive 123/132 sites with the formation of one F-frame.

The presented approach does not yield direct information on the relative arrangement of individual frames along the satellite DNA: do each of these frames repeat tandemly or is there a transition from one frame to another? If it is assumed that the length of the nucleosomal DNA is 188 bp (at a repeat length of 172 bp in the stDNA) the tandem arrangement of the same frames will hardly be possible.

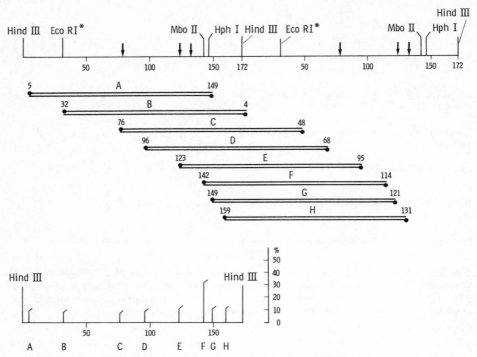

**Fig. 64.** Localization and relative abundance of eight nucleosome frames on AGM α-stDNA (Zhang et al. 1983). Two α-stDNA tandem repeats with numeration and localization of restriction sites are shown in the *top row*. *Arrows* indicate preferred micrococcal nuclease cleavage sites. Eight positions of the nucleosome frames (A–H) are shown underneath. The positions of the first and last nucleotide of the nucleosome core are indicated. The percentage of each frame is given at the bottom

Wu et al. (1983) have examined the structural organization of a defined part of AGM satellite chromatin. As indicated in the chapter devoted to structural features, about 15% of the monkey α-stDNA contains *Eco*RI sites which are clustered. These authors treated isolated nuclei with *Eco*RI, solubilized and isolated part of the satellite chromatin in a pure form and called it the *Eco*RI subset. This fraction revealed oligonucleosomes of different lengths, one of every four *Eco*RI-solubilized particles being a mononucleosome.

*Eco*RI-solubilized α-oligonucleosomes were fully resistant to subsequnt redigestion with *Eco*RI despite the presence of one *Eco*RI site in each α-stDNA 172 bp repeat. In the *Eco*RI-solubilized oligonucleosomes the only *Hind*III digested site is localized at a distance of 30 bp from one of the ends. The other internal *Hind*III sites are wholly resistant to enzyme digestion.

The obtained results demonstrate the specific arrangement of nucleosomal core particles in relation to the nucleotide sequence of the α-stDNA repeat in the *Eco*RI subset. The center of core particles is at a distance of 20 bp from the nearest *Eco*RI site and 50 bp from the *Hind*III site.

Strauss and Varshavsky (1984) isolated HMG-like α-stDNA binding protein (α-protein) from crude extracts of African green monkey cells. Three specific binding sites per α-stDNA repeat were detected. Sites II and III contain inverted 7 bp repeats of each other (AAATATC and GATATTT), while site I contains a weakly homologous sequence −TTAATTC.

The distance between outward borders of sites II and III is about 145 bp. Site I is positioned in the middle between sites II and III.

The equilibrium association constant of α-protein and α-stDNA is $\sim 5 \times 10^{10} \, M^{-1}$.

## 7.4 *Drosophila* Chromatin

The presence of a phosphorylated form of histone H1 is a characteristic feature of stDNA-containing nucleosomes from *Drosophila*. Blumenfeld et al. (1978) made a comparative analysis of two *Drosophila* species, *D.virilis* and *D. novamexicana*, differing in stDNA sets. *D.virilis*, with three stDNAs, contains five subfractions of histone H1; *D.novamexicana*, with a single stDNA, reveals only three such subfractions. Analysis of the histone subfractions, correlating with stDNAs, has shown that in vitro only histones H1 predominantly bind with *D.virilis* stDNAs.

Later the same group found that phosphorylated histone H1 accounts for less than 10% in *D.virilis* polytene cells with a stDNA content reduced to less than 1% due to underreplication (Billings et al. 1979). On the other hand, almost half of the histones H1 in brain diploid cells are phosphorylated.

Phosphorylated histone H1 was found also in *D.melanogaster* (Blumenfeld 1979). In different *D.melanogaster* tissues it composed 30−40% of histone H1. It has been suggested that the compactization of heterochromatin can be determined by the specific binding of the phosphorylated form of histone H1 with stDNA.

Hsieh and Brutlag (1979b), using hybridization on filters, partially purified *D.melanogaster* embryo protein, which prevalently binds with stDNA 1.688. The stDNA-protein complex formation proceeds at ionic strength and temperature corresponding to physiological conditions. The protein recognizes double-stranded DNA. A hybrid plasmid DNA, representing a pBR322 vector with an inserted 359 bp fragment of stDNA 1.688 was used for its binding. Formation of the complex requires the presence of DNA in a supercoiled form. The protein interacts with a limited region of the 359 bp repeat stDNA unit. The chemical nature of this protein has not been established.

A nonhistone D1 protein, rich in basic and acidic amino acids, with a molecular mass of 50,000 daltons, was detected in *D.melanogaster* nucleosomes containing stDNA 1.672 and 1.688 (Levinger and Varshavsky 1982). It should be noticed that the D1 and the protein described by Hsieh and Brutlag are two different proteins. The molar content of the D1 protein is about 10% of the amount of histone H1. The D1 protein is revealed in satellite DNA-containing nucleosomes in addition to the core histones and not instead of them. The stoichiometry of the D1-nucleosome complex has not been established precisely, it is assumed that there is one molecule per particle.

It has been shown that in vitro the D1 protein predominantly binds with AT-rich double-stranded DNAs. Though both stDNAs have an affinity for the D1 protein, stDNA 1.688 is incapable of competing with stDNA 1.672 even at a 100-fold excess.

Synthetic polynucleotides poly [d (A − T)] · poly [d (A − T)], poly (dA) · poly (dT), and stDNA 1.672 bind with an equal efficiency with protein D1. The separate synthetic polymer chains do not compete with stDNA.

The characteristic feature of *D.melanogaster* satellite chromatin is, besides the presence of protein D1, the absence of the ubiquitin H2A semihistone, a covalent conjugate of histone H2A and the protein ubiquitin. It is detected only in transcribing chromatin (Varshavsky et al. 1983).

## 7.5 Mouse Chromatin

Mouse satellite chromatin is readily isolated from total chromatin, as the repeating sequence of mouse stDNA does not contain restriction sites of some endonucleases (*EcoRI, AluI, BspI*) which extensively digest bulk chromatin. Using these enzymes, several laboratories have fractionated mouse chromatin into satellite and bulk chromatin and analyzed their protein components.

It was shown that chromatin, containing both stDNA and the main component, is characterized by the same protein: DNA ratio. The procedure of cross-linking of chromatin components with aldehydes, permitting to distinguish the proteins tightly bound and associated with chromatin, also did not reveal any difference between them (Mazrimas et al. 1979). Both types of mouse chromatin contained all the five major types of histones and a minor lysine-rich histone H1$^0$ which were observed in the initial undigested chromatin as well.

Other groups also found no difference between satellite and bulk chromatin in the content both of major and minor histones (Bernstine 1978; Zhang and Hörz 1982; Pashev et al. 1983). Only the absence in satellite chromatin of HMG proteins characteristic of expressible chromatin was noted (Zhang and Hörz 1982). Hyperacetylated forms of histones, found in bulk chromatin, were not observed in satellite chromatin fractions (Pashev et al. 1983).

Gottesfeld and Melton (1978) demonstrated that the nucleosomal repeat in mouse satellite chromatin consists of 195 ± 5 bp and is not distinguished from bulk chromatin.

In a later study Zhang and Hörz (1983) determined the arrangement of nucleosome frames along the mouse stDNA chain.

The authors identified the presence of 16 nucleosomal frames, with none being a dominating one (Fig. 65). There is no simple phase correlation between the stDNA repeats (234 bp) and nucleosomal periodicity, as the length of nucleosomal DNA in mouse satellite chromatin is 195−200 bp. This means that the tandem arrangement of any frame is hardly probable; nucleosomal positions must change constantly. Zhang and Hörz found that a correlation is observed between the arrangement of nucleosomes along the stDNA chain and the 9 bp of the basic internal repetition in mouse stDNA.

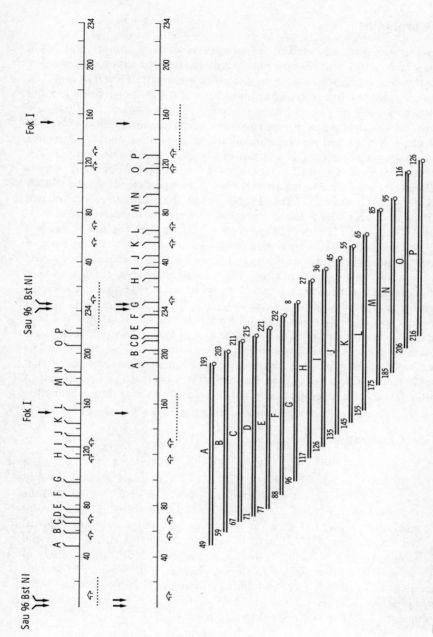

**Fig. 65.** Localization of 16 nucleosome frames on mouse stDNA (Zhang and Hörz 1984). The *top row* shows two stDNA tandem repeats and restriction sites. *Open arrows* indicate preferred cleavage sites of micrococcal nuclease. The start ( ⌐ , *top row*) and end ( ⌐ , *second row*) of the 16 nucleosome frames. Positions of the frames are shown underneath with the *numbers* referring to the start and end

## 7.6  Rat Chromatin

This chromatin was detailedly studied in Zachau's laboratory (Omori et al. 1980; Igo-Kemenes et al. 1980). Rat satellite chromatin was isolated in a highly purified form by treating cell nuclei with *Eco*RI restriction endonuclease. The DNA length in nucleosomes of satellite chromatin and in bulk chromatin was shown to be 185 and 195 bp, respectively.

Böck et al. (1984) determined the arrangement of nucleosomal frames along the rat stDNA chain. The procedures were mainly the same as those used to determine nucleosomal frames along the stDNA chains of the green monkey and mouse. The novelty of their study is that satellite fragments were extracted from rat total core DNA by cloning, sequenced, and the arrangement of core particles along the stDNA chain were identified. It was shown that 35−50% of all the nucleosomes occupy two main positions, the remainder is found in 16 less preferred positions (Fig. 66).

The rat stDNA repeat length (370 bp) is twice greater than the size of satellite chromatin nucleosomal DNA. Proceeding from this, it was assumed that pairs of nucleosomes or more complex, but distinct oligonucleosome configurations would give rise to "islands" or nucleosome arrays. The transition from one configuration to another may be due to deletion or insertion of the subrepeat 92/93 bp which interrupts tandem repetition regularity of the 370 bp unit.

As concerns histone composition, no difference between the rat satellite and bulk chromatin is observed (Omori et al. 1980; Mathew et al. 1981). Only a 50% decrease of the histone H1 content in satellite chromatin was noted and explained by its loss in the process of chromatin isolation. Neither did enrichment of the satellite chromatin with nonhistone proteins occur; on the contrary, a decrease in the content of HMG14 and HMG17 proteins was observed in comparison with bulk chromatin.

## 7.7  Bovine Chromatin

Bovine chromatin fragments have been isolated containing stDNA 1.715 (Weber and Cole 1982a). Isolated nuclei were treated with *Eco*RI restriction endonuclease to produce satellite chromatin. It was solubilized by lysis at low ionic strength and the chromatin fragments containing stDNA were isolated by chromatography on a malachite green resin column. The stDNA 1.715 content in the isolated fragments reached 70−90%. The nucleosomal DNA length in satellite chromatin and in the initial soluble chromatin was 186 ± 7 and 193 ± 5 bp, respectively. A distribution study of *Eco*RI sites in satellite chromatin led to the conclusion of the absence of specific nucleosome phasing in stDNA 1.715 containing chromatin. No difference in protein content was observed between satellite and total chromatin with the exception of a 30% less content of histone H1 in satellite chromatin (Weber and cole 1982b).

## 7.8  Specific Binding of stDNA with Microtubular Proteins

A study by Wishe et al. (1978), though not concerned directly with heterochromatin structure, showed a specific "recognition" of stDNAs by proteins, an event which can occur also in heterochromatin.

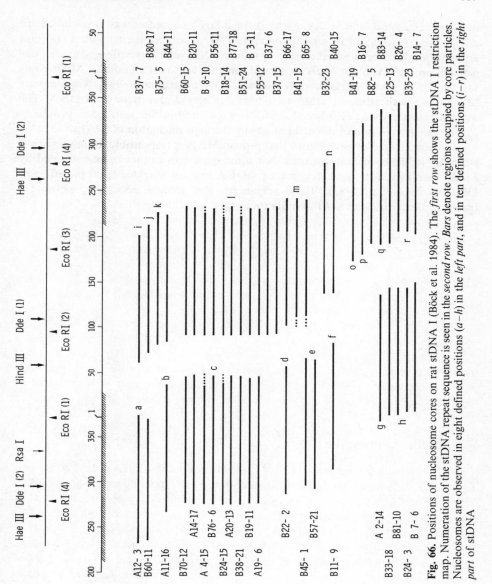

**Fig. 66.** Positions of nucleosome cores on rat stDNA I (Böck et al. 1984). The *first row* shows the stDNA I restriction map. Numeration of the stDNA repeat sequence is seen in the *second row*. *Bars* denote regions occupied by core particles. Nucleosomes are observed in eight defined positions (*a*–*h*) in the *left part*, and in ten defined positions (*i*–*r*) in the *right part* of stDNA

The centromere is known to be the part of chromosomes interacting with spindle fibers during meiosis and mitosis. However, up to the present little data are available on the nature of centromere components and the mechanism of their interaction with spindle microtubules. The authors fractionated microtubular proteins from hog brain to obtain tubulin and also the so-called microtubular-associated proteins (MAPs). Tubulin alone was shown to be incapable of binding with DNA, but binding occurs in the presence of MAPs. The question of whether MAP binds with DNA uniformly or does preferential binding occur with defined DNA sequences was also studied. To clarify this question, an increasing amount of MAP was incubated with a constant amount of a $^3$H-labeled satellite or total mouse DNA, and the resulting complex was isolated on nitrocellulose filters. It turned out that an equal amount of MAP binds satellite DNA twice more effectively than total DNA. An unlabeled stDNA competes for binding with labeled stDNA to a much greater degree than total DNA. The obtained data provide evidence that stDNAs localized in the centromere proximity can be the in vivo microtubular binding site through mediation of MAPs.

Subsequent studies have shown that protein $MAP_2$ is responsible for binding with stDNAs. It has been demonstrated that some sequences in the complex formed by $MAP_2$ with the 234 bp repeating mouse stDNA unit are shielded from the effect of DNase (Avila et al. 1983). These sequences, representing $poly(dA)_4 \cdot poly(dT)_4$ tracts, evidently are the $MAP_2$-binding regions.

# 8 What Unites stDNAs?

It is evident from the review of the literature on the structure of stDNAs that they have different nucleotide sequences and occur in different amounts in the genome. There are both light AT-rich, and heavy GC-rich satellites. However, despite species specificity, stDNAs seem to have one common feature – the capability to participate in the formation of constitutive heterochromatin.

What are the structural features of stDNAs facilitating the formation of constitutive heterochromatin? Can any physico-chemical properties be found uniting different stDNAs?

## 8.1 Specific Conformation?

An effort was made (Selsing et al. 1976) to clarify whether stDNA is characterized by a conformation different from the classical A and B forms. It is known that the polynucleotide duplex $poly[d(A - T)] \cdot poly[d(A - T)]$ is characterized by a D-conformation. A study was made of AT-rich stDNAs, the *Gecarcinus lateralis* light stDNA ($\varrho = 1.677\,g/cm^3$), the *D.ordii* stDNA ($\varrho = 1.671\,g/cm^3$) and the mouse stDNA ($\varrho = 1.690\,g/cm^3$). An X-ray structural analysis of these DNAs showed that they are characterized only by the orthodoxal geometry of the A and B forms.

Reports have appeared recently that sequences capable of forming the Z form of DNA have been encountered in stDNAs (Skinner et al. 1983a). These data, as yet, refer only to *Gecarcinus lateralis* stDNA, and any generalization in this direction would be premature.

## 8.2 Length of the Repeating Unit?

In most cases it essentially varies in different stDNAs. As mentioned above, there are "simple" and "complex" stDNAs with basic repeat lengths differing by one or more orders of magnitude. The repeat length both of simple and complex stDNAs varies in a rather wide range. There is an interesting suggestion that there could be a common repeat in animal stDNAs with a size of about 170 bp, the presence of which is determined by the structure of constitutive heterochromatin nucleosomes (Maio et al. 1977; Musich et al. 1978). However, up to the present this hypothesis has not been confirmed.

## 8.3 Base Sequence?

No, the primary structures of repeating elements differ in different stDNAs, and in some cases even within one species (*D.melanogaster*, *D.ordii*, etc.). However, it should be mentioned at the same time than individual stDNAs within such a systematic large group as animals may have a common origin.

Salser et al. (1976) were the first to propose a hypothesis on the existence of a common "library" of satellite sequences in animals. They showed that *D.ordii* stDNA contains essentially the same sequence as the guinea pig α-stDNA. It was thought that rapid evolutionary changes involve only the quantitative content of these sequences in the genome. On this basis a suggestion was made that rodents (and, probably, other animals) may possess a common library of satellite sequences at a low level (each sequence in a quantitiy less than requisite for detection in the density gradient). According to this model, the rapid evolutionary variability undergone by stDNA is mostly of a quantitative nature. These quantitative changes may bring about a multiple increase or decrease of the relative content of a defined stDNA. However, an assumption is permitted that the disappearance of stDNAs in most cases is only a seeming one; in reality the stDNA drops below the level of their detection as density stDNAs, i.e., down to the level of the satellite library. In other words, the appearance of a "new" stDNA in most cases does not occur de novo as suggested in other theories, but on the basis of amplification of one of the satellites present in the library at a low quantitative level.

According to this theory, "candidates" for stDNA appear frequently, but they rarely acquire biological functions to warrant their conservation in the library for a long period of evolution.

A comparative analysis was performed of stDNAs from four rodent species belonging to two suborders of *Rodentia*: kangaroo rat (HS-α stDNA), pocket gopher, guinea pig (α-stDNA), and antelope ground squirrel (Mazrimas and Hatch 1977; Fry and Salser 1977). The stDNAs compared displayed close values of $T_m$, base compositions, and buoyant densities of separate stDNA strands in the alkaline CsCl gradient. Pyrimidine oligonucleotides of individual strands as well as fragments produced by RNase $T_2$ digestion of the ribo-G-substituted satellite DNA showed a significant coincidence (Table 20). In all four stDNAs studies not only the major, but the variant sequences also were identical. All of these sequences have been conserved during a prolonged period of evolution. It is assumed that divergence of the examined rodents occurred $50 \cdot 10^6$ years ago.

Many hypotheses on the functional role of stDNAs proceed from the fact that stDNAs are capable of specific interaction with "recognition" proteins. It follows that the satellite sequence library must be accompanied by a library of genes coding for the recognition proteins. In such a case, one should assume that it is, namely, the recognizing proteins that determine the significant conservation of satellite sequences.

The library theory was substantially upheld when the primary structure of stDNA repeating units was defined for several animals. Thus, Pech et al. (1979a) observed a homology between the sequences of rat stDNA I and mouse stDNA. A defined homology is observed also with stDNAs from other species (Fig. 67). According to the authors, this homology can, in the first place, demonstrate to some degree the evolutionary relationship of species. Secondly, it can also be assumed that short sequences

**Table 20.** Percentage of major fragments produced by RNase T$_2$ digestion of the rG-substituted H-strand of stDNAs (Fry and Salser 1977)

| Sequence | Kangaroo rat | Pocket gopher | Guinea pig | Antelope ground squirrel |
|---|---|---|---|---|
| rGp | 46 | 33 | 31 | 29 |
| T−T−A−rGp | 23 | 25 | 27 | 23 |
| T−T−T−A−rGp | 13 | 13 | 24 | 26 |
| (T$_2$, A$_2$)rGp | 3.9 | 5.8 | 3.7 | 5.8 |
| T−rGp | 2.9 | 5.7 | 4.2 | 5.1 |
| (T$_2$, A,C)rGp | 2.8 | 2.4 | 3.0 | 1.7 |
| (C, T)rGp | 2.3 | 2.8 | 1.0 | 0.5 |
| A−rGp | 1.6 | 2.6 | 1.2 | 1.8 |
| T$_3$−rGp | 1.0 | 2.2 | 1.4 | 2.3 |
| (T$_2$−C)rGp | 0.8 | 1.9 | 0.7 | 0.8 |
| (A, T)rGp | 0.6 | 1.9 | 0 | 0.4 |
| T−T−rGp | 0.6 | 2.2 | 0.5 | 1.3 |
| C−rGp | 0.5 | 0.3 | 0.2 | 0.4 |
| (C, A, T)rGp | 0.5 | 1.1 | 0.5 | 0.7 |
| (T, A$_2$)rGp | 0.3 | 0.7 | 1.0 | 0.8 |

detected in various stDNAs play a role in binding specific proteins, and, consequently, possess a functional significance, i.e., to determine the stDNA capability for compactization and participate in the process of heterochromatin formation.

To what extent does the library theory apply to other groups of organisms? Do plants have a common library? Is the role in compactization of heterochromatin intrinsic to one particular sequence, or are there several such sequences? Only additional experiments will resolve these questions.

## 8.4 5-methylcytosine?

It has been suggested in the literature that different stDNAs can be united by the presence in them of m$^5$C (Musich et al. 1977b). The authors note that constitutive heterochromatin can contain both GC- and AT-rich sequences. Hence, it is hard to imagine any histone or nonhistone protein structure that could react stereospecifically with such highly variable sequences. Musich et al. proposed a hypothesis according to which stereospecificity can be defined by the minor base m$^5$C. The frequency of m$^5$C occurrence in eukaryotic DNAs is higher in satellite or hrDNA than in the other genome fractions. The authors assume that a nonspecific hydrophobic interaction of m$^5$C probably occurs with the nonpolar regions of one or several chromosomal proteins. Treatment of animal chromatin with micrococcal nuclease under conditions of 50 % cleavage showed that 75 % of m$^5$C is protected against digestion.Subsequent studies, however, reported that in some stDNAs cytosine is not methylated. Thus, m$^5$C in *D.melanogaster* stDNA 1.688 is absent (Hsieh and Brutlag 1979a).

Moreover, this minor base is not detected at all in the *D.melanogaster* genome (Mazin et al. 1984). There are also data that m$^5$C is absent or is found in a small quantitiy in other stDNAs (Beridze 1980b; Wagner and Capesius 1981; Streeck et al.

I    5' ... CAAATTTTGA   TTA...   3'

II    5' ... ATGGCGA   GAAAACTGCA   AAACATGAAA   TCAG......C   ACTGACGACT   GAAAATGACG   AAA ... AAA   ACGTGBAA...   3'

III   5' ... CAAATTTTGA   TAAATCTTTA   AAAGTACACA   TATTACAAGA   GCAGGCTACT   TGAATTCACA   GAGAAACAGT   GTTTCAGTTC   GTTA...   3'

IV    5' ... GA   TCACGTGACT   GATCATGAAC     ...   3'

V     5' ...CACA   GAGCAGACTT   GAAACACTCT   TTTT...   3'

VI    5' ...CACA   GAGTTACATC   TTTCCCTTCA   AGAA...   3'

**Fig. 67.** Homology between sequences of rat stDNA I (*III*), stDNAs of *D.melanogaster* (*I*), mouse (*II*), calf (*IV*), human (*V*) and African green monkey (*VI*) (Pech et al. 1979a). The following sequences are presented: a part of *D.melanogaster* stDNA 1.688, an incomplete sequence of mouse stDNA, the rat stDNA sequence from position 310 to position 34, the prototype sequence of segments A and C of bovine stDNA 1.706, human alphoid stDNA (positions 61–88), and AGM α-stDNA (positions 39–66)

1982). Consequently, it would be ungrounded to unite stDNAs according to their $m^5C$ content.

## 8.5 Concluding Remarks

As evident from the material considered in this chapter, the question remains open as to what structural features lend stDNA the capability to participate in the formation of constitutive heterochromatin, one of the essential points in the problem of satellite DNAs. It does not seem possible to give a conclusive answer to the posed question proceeding from the experimental data available at present.

In my opinion, the primary structure of stDNAs must contain the key for understanding the compact state of heterochromatin.

It can be thought that a common feature of stDNAs is a regular distribution of purine (whether be it A or G) and pyrimidine bases (C or T) along their chains. These regularities, if they exist, can be revealed by computer analysis of known stDNA repeat unit sequences. Such regular tracts can impart a specific "tertiary" structure to stDNAs which, in turn, determine the compact structure of constitutive heterochromatin.

What experimental data are available to support the proposed viewpoint?

Studies carried out in recent years have shown that DNA molecules have a significant conformational flexibility. The DNA helical configuration is not an absolutely regular one. It exhibits geometrical irregularities which reflect the base sequence along the DNA chain. It has been shown that fragments of *Leishmania tarentolae* kinetoplast DNA during polyacrylamide gel electrophoresis reveal unexpected properties. The anomalous migration rate, characteristic of these fragments, approaches the normal with an increase of the gel pore size. This property is explained by the presence of curvatures in the DNA molecule (Marini et al. 1982; Hagerman 1984).

It has been suggested that a purine clash is responsible for the curvature observed in kinetoplast DNA (Hagerman 1984). Purine clash is understood as steric repulsive forces between purine bases in consecutive base pairs, but on opposite chains (Calladine 1982).

According to Wu and Crothers (1984), $CA_{5-6}T$ tracts periodically repeated at 10 bp intervals are responsible for sequence-directed bending of the kinetoplast DNA fragments.

Theoretical calculations carried out for tetramers with a different base sequence were used to predict the form of DNA molecules with a known sequence (Ulyanov and Zhurkin 1984).

A precise determination of the electrophoretic mobility of stDNA fragments in various conditions as well as the theoretical calculation of their form can provide an answer as to the validity of my assumption.

# 9 Origin

## 9.1 Saltatory Replication

The first scheme of origin and evolution of hrDNAs and, in particular, of stDNAs, was proposed by Britten and Kohne (1966, 1968). According to the authors, the formation of stDNA was a result of a sudden event which they called saltatory replication.

Several steps are distinguished in this process: (1) the sequence undergoes manifold replications; (2) the formed copies are integrated into the chromosome; (3) the satellite sequence associates with a favorable genetic element; (4) as a result of natural selection the satellite sequence becomes a component of the entire species.

The authors believe that each step must be characterized by a low probability of an event. In principle, saltatory replication can be observed in an individual organism by analogy with mutation. It is not excluded that saltatory replication in the future may be induced artificially. Figure 68 shows Britten and Kohne's scheme depicting the formation and evolution of repeating sequences.

StDNA formation on the basis of saltatory replication could presumably proceed via two mechanisms. The first one is the "slippage" mechanism in the DNA-polymerase reaction, permitting a multiple copying of a short repeat (Kornberg et al. 1964; Wells et al. 1967). Another possible mechanism of the considered phenomenon is the "rolling circle" mechanism participating in amplification of ribosomal RNA genes in *Xenopus laevis* oocytes (Hourcad et al. 1973). However, a defined length of a chain is required to form circular DNA. Hence, a short sequence must be preliminarily amplified to a length sufficient for closing the circle.

There is evidence that quantitative changes in heterochromatin and, consequently, in stDNAs, can be induced by a chromosome break or structural rearrangements.

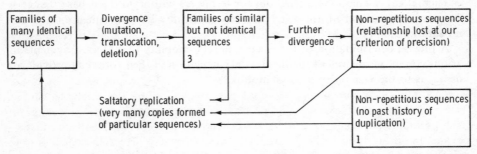

**Fig. 68.** Formation and evolution of the eukaryotic DNA repeating sequences (Britten and Kohne 1966)

Thus, the formation of ruptures in the heterochromatin itself or in its proximity enhances saltatory changes of its sizes (John and Miklos 1979).

## 9.2 Unequal Crossing-Over

A completely different mechanism of repetitive DNA formation, based on unequal crossing-over, was suggested by Smith (1974, 1976, 1978). It is common knowledge that crossing-over results in two recombinant molecules which receive the left part from one initial molecule and the right part from the other one. In many crossing-overs the alignment of DNA molecules proceeds faultlessly, with a colinear relationship of base pairs in the initial and recombinant sequences. In some crossing-overs, however, the participating molecules are aligned in a staggered manner and form recombinant molecules longer or shorter than the initial ones. In such a case, it is an unequal crossing-over (Fig. 69).

A tandem duplication is formed in the long recombinant with a length equal to that of noncoincidence, while the short recombinant contains a deletion with the same sequence.

Unequal crossing-over occurring in the germ-line cells during meiosis or mitosis increases the number of deletions and tandem duplications which are inherited by following generations and are definitely of evolutionary importance. According to this theory, the accumulation of many such deletions and tandem duplications during a specific period of time results in the formation of repetitive DNAs (Fig. 70).

Computer analysis has shown that if unequal crossing-over occurs in an initial DNA constantly and in a random fashion, such a DNA becomes a chain of tandem repeats.

The frequency of deletions and duplications, just as that of other usual mutations, undergoes a genetic drift. The genetic drift removes most of the deletions and duplications from the population and simultaneously increases the frequency of some deletions and duplications, and, as a result, leads to their spreading throughout the whole population. Consequently, according to the theory of unequal crossing-over, a repeating organization is a characteristic state of DNA, if natural selection does not impede the random character of unequal crossing-over.

Smith's hypothesis on the role of unequal crossing-over in tandem repeat amplification must be coordinated with the genetic processes taking place in the cell. When does unequal crossing-over occur, during meiosis or mitosis of germ-line cells? Is there any experimental proof demonstrating accumulation or deletion of hrDNAs through unequal crossing-over? Is the frequency of unequal crossing-over sufficient for hrDNA formation?

The quantitative variability of constitutive heterochromatin in individuals of one and the same species, known as heteromorphism, has been demonstrated by up-to-date methods of chromosome studies and, in particular, by the C-banding technique. The variability of constitutive heterochromatin is accompanied by a change of hrDNA and, in particular, of stDNA. No crossing-over occurs in constitutive heterochromatin during meiosis. Rearrangements of constitutive heterochromatin are observed only during mitosis. The only possibility for inherited changes in the constitutive heterochromatin is an unequal exchange between sister chromatids during

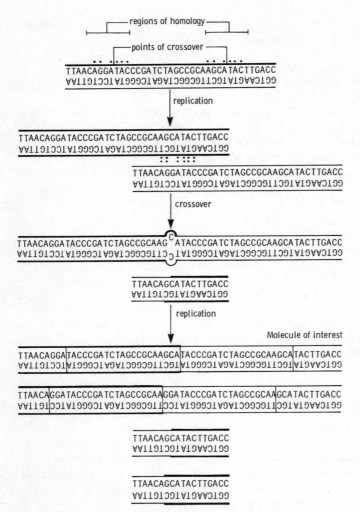

**Fig. 69.** Unequal crossing-over between sister chromatids (Smith 1976). DNA chains formed from the upper strand of the starting molecule are indicated by *bold lines*; those formed from the lower strand by *light lines*. For unequal crossing-over to occur, the initial molecule must contain a repeat or an inverted repeat. Four molecules are formed after crossing-over and subsequent replication, of these, two final long molecules contain a tandem duplication of the starting sequence. The duplicated sequences are denoted by a *vertical line*

premeiotic mitoses or else during replicative division of meiosis (John and Miklos 1979).

## 9.3 Two-Step Mechanism Formation

Walker (1971) suggested a two-step scheme of stDNA origin and evolution (Fig. 71). Walker based his reasoning on the experimental data of Southern (1970) on the

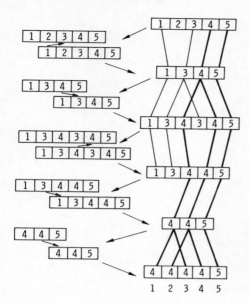

**Fig. 70.** Fixation of the tandem duplication process on the basis of unequal crossing-over (Smith 1974). Each *numbered box* corresponds to a defined repeat. *Bold lines* join repeats observed in the final sequence

structural organization of guinea pig α-stDNA demonstrating the existence of a stepwise process of stDNA formation. Walker also took into consideration in situ hybridization data indicating the predominant localization of stDNAs in the centromeric heterochromatin (Pardue and Gall 1970).

According to Walker's concepts, the predecessor of stDNA is a spacer region located between the genes of 18S and 28S rRNA, while other sources (Yunis and Yasmineh 1971) assume that it is a small segment of pericentric or perinucleolar DNA. All the authors up to the present are in agreement that formation of the majority of simple stDNAs proceeded in two stages: the first was the formation of a relatively long segment (about 200 bp in length) on the basis of a short basic sequence; the second was the appearance of a highly repeating unit by amplifiation of the long segment. As concerns complex stDNAs, they were formed by direct amplification of long segments lacking internal repeats, omitting the first stage. It is not excluded, however, that a thorough computer analysis can reveal internal repeating elements in comlex stDNAs as reported in the case of bovine stDNA 1.715 (Plucienniczak et al. 1982).

The first stage, the formation of a 200 bp segment, can be called a microprocess in chromosomal DNA evolution; it is quantitatively insignificant. The second stage, a multithousandfold amplification of the long segment, is a macroprocess resulting in the formation, in a number of cases, of material composing a significant part of the genome. It may be that the mechanisms of these two stages are unlike. Firstly, regions differing 10–20 times in length are repeated; secondly, the number of repeating steps accompanying the formation of these two levels of stDNA organization also markedly differ.

Southern (1975) in suggesting a scheme of stDNA evolution assumed that a mechanism of saltatory replication functions at the first stage of short stDNA se-

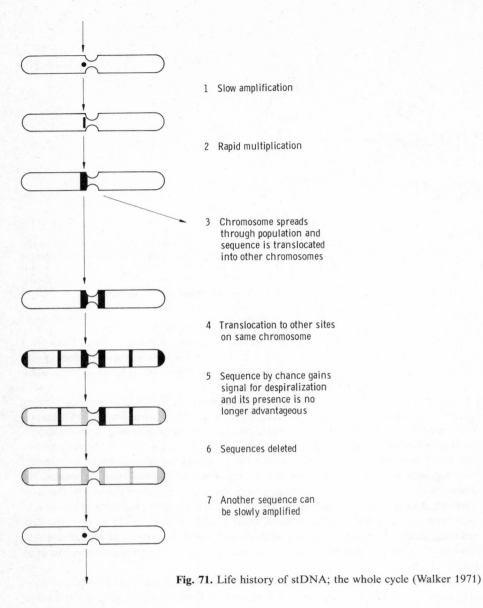

1 Slow amplification

2 Rapid multiplication

3 Chromosome spreads through population and sequence is translocated into other chromosomes

4 Translocation to other sites on same chromosome

5 Sequence by chance gains signal for despiralization and its presence is no longer advantageous

6 Sequences deleted

7 Another sequence can be slowly amplified

**Fig. 71.** Life history of stDNA; the whole cycle (Walker 1971)

quence formation and that the mechanism of unequal crossing-over can explain the staggered register of long-range periodicity formed at the second stage of evolution.

Musich et al. (1978) presented a very attractive theoretical conception on the formation of long-range periodicity of stDNA. The authors suggested that the second stage of stDNA formation is determined by the subunit structure of chromatin. Preference was given to the mechanism of unequal crossing-over as the one determining the given stage. In the scheme (Fig. 72) it is assumed that crossing-over occurs

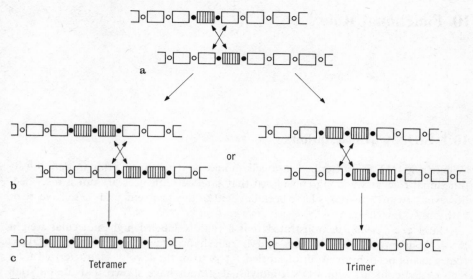

**Fig. 72.** Hypothetical scheme of repeating sequence expansion through unequal crossing-over at the nucleosome junctions (Musich et al. 1978). **a** Crossing-over between mononucleosomes; **b** crossing-over between dinucleosomes. At end-to-end crossing-over a geometrical expansion is observed (**c** left). Crossing-over with the participation of one nucleosome results in an arithmetical expansion (**c** right)

between nucleosomes, at the junctions of which there are direct or inverted short repeats requisite for recombination. The chain length will increase rapidly as a result of a number of subsequent recombinations. The recombinations can form blocks producing both an arithmetical series of oligomers (i. e., a monomer, dimer, trimer, etc.) and a geometrical series (a monomer, dimer, tetramer, octamer, etc.). It should be noted that both types of organization were detected in stDNAs by restriction analysis.

The above mentioned hypothesis has a rational kernel, since the consideration of amplification mechanisms exclusively from the standpoint of only the DNA structure, without taking into account that the amplified DNA is organized in the form of chromatin, is to some degree a mechanistic approach to the problem, disregarding the actual situation.

The detection of insertion sequences in the repeating units of several stDNAs indicates the specific mechanism of formation of these stDNAs: the insertion of a DNA fragment into the repeating sequence and the subsequent amplification of the region containing the insertion sequence and part of the repeat (Streeck 1981).

As mentioned above, complex stDNA repeat units have been shown to contain variable and conserved regions, in particular, bovine stDNA 1.715 and *D.melanogaster* stDNA 1.688 (Hsieh and Brutlag 1979a; Roizes and Pages 1982). It has been suggested that the mechanism of gene conversion is responsible for maintaining the homogeneity of the conserved regions in the repeat units of stDNAs (Roizes and Pages 1982).

# 10 Functional Role

## 10.1 StDNA Transcription

Whether stDNA is transcribed in a cell should be determined prior to defining its functional role. It was long considered that satellite sequences are not transcribed, however, few investigations have been devoted to this problem, and these have dealt with only a few stDNAs.

Harel et al. (1968) demonstrated that a rapidly labeled high molecular weight RNA from mouse liver, kidney, and L cells hybridize with $^{32}$P-labeled stDNA. Experiments on hybridization of labeled RNA from the L-cells, characterized by a high uracyl content, with stDNA individual strands have shown that the GC-rich L-chain is the one transcribed.

Walker (1971) in a critical review of the work noted that Harel's data demonstrate that only 1% of the stDNA is hybridized with the RNA. Walker assumed that this could be due to a minor transcribed component of a satellite fraction which is covalently linked to the stDNA.

Flamm et al. (1969) reported that less than 1 of 60,000 RNA molecules from different mouse organs can be complementary to a stDNA (Table 21).

An insignificant hybridization of total RNA from HeLa cells with human stDNAs has been observed by other authors who assumed that even if stDNAs are transcribed, the transcripts compose an extremely small part of the total cellular RNA (Melli et al. 1975) (Table 22).

In analyzing stDNA transcriptability, it should be kept in mind that satellite-like sequences in many species are interspersed with the main component (Stambrook 1981). The detection of transcripts can in reality mean the transcription of satellite-like sequences interspersed in the main component and not that of the stDNA itself.

Some recently published papers have reported that stDNA transcription actually takes place in some organisms.

**Table 21.** Hybridization of a 10,000-fold excess of RNA with single strands of mouse stDNA (Flamm et al. 1969)

| $32_{P-DNA}$ $(2 \cdot 10^{-5}$ mg/ml) | Time of incuba-tion, min | % of hybridized RNA $(2 \cdot 10^{-1}$ mg/ml)[a] | | | |
|---|---|---|---|---|---|
| | | Control | Liver | Spleen | Kidney |
| H-Strand | 0 | 0.2 | 2.2 | 1.5 | 0.6 |
| | 60 | 1.1 | 5.6 | 5.0 | 3.3 |
| L-Strand | 0 | 0.6 | 1.5 | 5.1 | 1.6 |
| | 60 | 1.4 | 5.0 | 10.4 | 2.2 |

[a] Estimation of % of hybridized RNA can be done by dividing the given values by $10^4$

**Table 22.** Hybridization of human cellular RNA to pure stDNAs (Melli et al. 1975)

| DNA + RNA | | % Counts RNA input hybridized |
|---|---|---|
| stDNA I + total RNA | | 0.0006 |
| stDNA II + total RNA | | 0.0006 |
| stDNA III + total RNA | | 0.0006 |
| stDNA I + hnRNA[a] | (a) | 0.011 |
| | (b) | 0.050 |
| stDNA III + hnRNA | (a) | 0.011 |
| | (b) | 0.020 |

[a] hnRNA – heterogeneous nuclear RNA

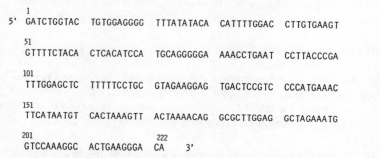

```
    1
5'  GATCTGGTAC  TGTGGAGGGG  TTTATATACA  CATTTTGGAC  CTTGTGAAGT

    51
    GTTTTCTACA  CTCACATCCA  TGCAGGGGGA  AAACCTGAAT  CCTTACCCGA

    101
    TTTGGAGCTC  TTTTTCCTGC  GTAGAAGGAG  TGACTCCGTC  CCCATGAAAC

    151
    TTCATAATGT  CACTAAAGTT  ACTAAAACAG  GCGCTTGGAG  GCTAGAAATG

    201                     222
    GTCCAAAGGC  ACTGAAGGGA  CA    3'
```

**Fig. 73.** Nucleotide sequence of *Notophthalmus viridescens* stDNA 1 (Diaz et al. 1981)

Varley et al. (1980) observed transcription of stDNA on lampbrush chromosomes from, the newt *Triturus cristatus carnifex*. The authors isolated and cloned a 330 bp segment of *T.c.carnifex* stDNA with the following cytological hybridization of the labeled segments with metaphase chromosomes. Autoradiograms displayed hybridization of labeled stDNA segments with RNA present in the *T.c.carnifex* loop region of the long arm of chromosome I. The observed effect disappeared completely if the metaphase chromosome preparation was preliminary incubated with an RNase.

The authors assume that the transcription of stDNA can be incidental and that RNA polymerase overrun normal termination signals and read right through stretches of the adjacent stDNA chain.

A more detailed study of this question was done by Diaz et al. (1981) using stDNA 1 of the newt *Notophthalmus viridescens*. StDNA 1 consists of 222 bp repeats (Fig. 73) and occurs both in centromeric heterochromatin of all chromosomes and in two noncentromeric regions at the sphere loci in lampbrush chromosomes. Hybridization in situ has shown that the stDNA 1 sequence is transcribed in the sphere loci, whereas in the centromeric regions it is inactive.

In *N.viridescens* five histone genes are clustered as a DNA segment 9 kb in size. Each cluster, consisting of five genes, is separated by a stDNA 1 segment. The size of most satellite segments is over 50 kb.

It is assumed that transcription initiates at the promoter of histone genes, fails to terminate at the end of the gene,and continues into adjacent satellite region. As a result, very long transcripts are formed containing both histone gene and stDNA sequences.

Diaz and coauthors have shown that both chains of *N.viridescens* stDNA 1 are transcribed. The explanation may be that in the gene cluster the sequences coding for histones H1, H3, H2A and H4 are localized on one DNA strand and those coding for H2B on the opposite.

Transcription of rat stDNA I was observed also in nuclei of hepatoma tissue culture cells (Sealy et al. 1981). The nuclear RNA molecules hybridized with stDNAs were larger than 16S. Control experiments did not reveal any nonspecific hybridization or stDNA admixtures in the RNA preparations. In general, the functions of stDNA transcripts are unknown. Not a single case of their translation has been described and up to the present their presence in the cytoplasm has not been revealed.

## 10.2 Role of stDNA in Somatic Cells

Several cases are known when stDNAs decrease sharply in quantity or are completely eliminated in somatic cells, while in germ-line cells their quantitiy and set remain constant. An assumption was made that stDNAs (at least some of them) do not have somatic functions, but play an essential role only in germ-line cells (John and Miklos 1979).

A part of the chromosomes of some ascarids are eliminated at the stage of early cleavage (this process is known as chromosome diminution), while germ-line chromosomes preserve their intactness. During diminution the chromosome heterochromatic regions disappear and the chromosomes remain euchromatic ones. Moritz and Roth (1976) made a comparative analysis of somatic and germ-line cells in two ascarid species. The somatic cells of both species reveal one component with a density of $1.700 \text{ g/cm}^3$; the germ-line cells of *Parascaris equorum* show two additional components ($\varrho = 1.692$ and $1.697 \text{ g/cm}^3$), and those of *Ascaris lumbricoides*, one additional component ($\varrho = 1.697 \text{ g/cm}^3$) (Table 23).

Chromatin diminution is accompanied by the disappearance of additional components (Fig. 74). Quantitatively, *P.equorum* and *A.lumbricoides* lose 85% and 22%

**Table 23.** DNAs of somatic and germ-line cells of *Ascaris* (after data of Moritz and Roth 1976)

| Species | Amount of DNA $10^{-12} \text{ g}$ | | Dimi-nution, % | Buoyant density of DNA, $\text{g/cm}^3$ | | |
|---|---|---|---|---|---|---|
| | Germ-line cells | Soma-tic cells | | Germ-line cells | | Somatic cells |
| | | | | Major compo-nent | Satellite | |
| *Parascaris equorum* | 1.2–2.1 | 0.25 | 85 | 1.700 | 1.692 1.697 | 1.700 |
| *Ascaris lum-bricoides* | 0.32 | 0.25 | 22 | 1.700 | 1.697 | 1.700 |

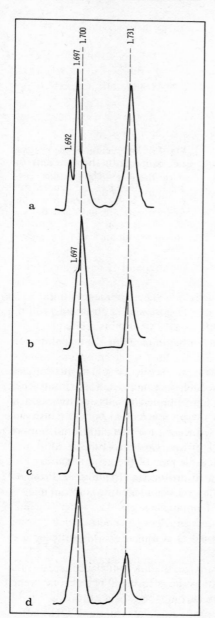

**Fig. 74.** Distribution of *Ascaris* DNAs in the CsCl density gradient (Moritz and Roth 1976). Spermatids: **a** *Parascaris equorum* (germ-line DNA) **b** *Ascaris lumbricoides*; larvae: **c** *P.equorum* (somatic DNA); **d** *A.lumbricoides*. 1.731 — *M.lyso-deikticus* DNA

DNA during diminution. Germ-line cells were shown to contain a stDNA which was eliminated during diminution (Roth 1979).

Though chromatin diminution removes more than 99.5% stDNA, some part of it still remains in somatic cells as tandemly arranged repeating units.

The nucleotide sequence of the uncloned repeating unit of eliminated stDNA of *A.lumbricoides* has been determined (Streeck et al. 1982). 70% of the stDNA consists

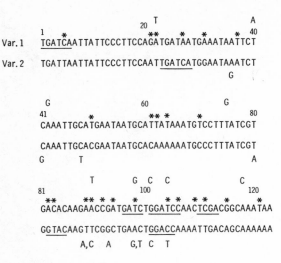

Fig. 75. Comparison of the nucleotide sequence of two variants of *Ascaris suum* stDNA (Streeck et al. 1982). Differences are denoted by *asterisks*, base substitutions in minor variants are indicated by *letters*. Restriction sites of some restriction endonucleases are *underlined*

of two main variants of a repeat 123 bp in size (Fig. 75). Their sequences indicate a high level of homology (base substitution in 24 of 123 positions is observed without clustering of the substitutions). The remaining 30 % stDNA contains related repeats. An important point is that cytosine is methylated neither in the CpG-dinucleotide nor in any other position.

Müller et al. (1982) cloned and determined the sequence of a number of *A.lumbricoides* stDNA fragments. The variants differ in small deletions, insertions, and base substitutions. An analysis based on the obtained results has revealed a 121 bp consensus sequence permitting to detect the presence of 11 bp short internal subrepeats ($GCA_A^TTT_G^TTGAT$). The consensus sequence of Müller and coauthors is close in structure with the sequence of variant 2 of Streeck et al. (1982).

Another example of elimination of a considerable part of the genetic material in the process of organism development is the ciliated protozoan *Stylonychia* (Prescott et al. 1973). Each *Stylonychia sp.* cell contains four genetically identical diploid micronuclei (germ-line nuclei) and two macronuclei (somatic nuclei). The macronucleus contains 600-fold more DNA than the diploid micronucleus. A macronucleus is one of the products of division of a synkaryon formed at conjugation. Micronuclei are other products of synkaryon division.

Before transformation into a macronucleus anlage, the micronucleus increases its DNA content 16-fold and a polytene-like chromosome is formed. The chromosome is cleaved soon after formation. Cleavage of the chromosomes is accompanied by a reduction of 93 % of the DNA to acid-soluble products. The remaining DNA after some time undergoes multiple replication and towards the end of the process reaches an amount 600 times exceeding that in the initial diploid micronucleus.

Four DNA components are observed in the CsCl density gradient of the micronucleus, while only one is detected in the macronucleus – three components are absent (Fig. 76). The absence of these components is due to the destruction of 93 % of the DNA in polytene-like chromosomes. The DNA fraction observed in the micronucleus and absent in the macronucleus is characterized by satellite-like properties.

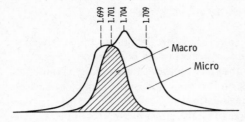

**Fig. 76.** Distribution of *Stylonichia* micro- and macronuclear DNA in the CsCl density gradient (Prescott et al. 1973). The DNA of micronuclei contain four components with densities of 1.699, 1.701, 1.704, and 1.709 g/cm$^3$. Macronuclear DNA consists of a single component with a density of 1.701 g/cm$^3$

*Stylonychia* studies have shown the significant reduction, if not elimination, of defined micronucleus fractions in the process of macronucleus formation.

Macronuclear DNA contains all the genetic information necessary to support cell growth and reproduction, as well as all the metabolic functions of a vegetative cell. More than 93% of RNA synthesis occurs in the macronucleus during vegetative growth. It follows that the three DNA components found only in the macronucleus are necessary solely to support germ-line cells.

The reduction of stDNA in the somatic cells is characteristic of a limited number of organisms, and it would be speculation to generalize the data and on these grounds fully exclude any stDNA somatic functions.

There are some data in the literature that could provide evidence for defined somatic functions of stDNAs. Ivanov and Markov (1980) demonstrated that stDNAs isolated from normal mouse tissues and from Ehrlich ascites tumor cells showed no distinctions under treatment with ten different restriction endonucleases or in the thermal stability of duplexes formed through cross-hybridization of separate stDNA strands from normal and tumor cells. Ehrlich ascites tumor was obtained in 1905 from a spontaneous tumor tissue of mouse mammary gland and converted into ascite form in 1932. During these years the cell cultivations underwent considerable genetic alterations resulting in various substrains with different karyotypes. Ivanov and Markov experimented with one of the strains which, in contrast to a normal karyo- type containing 40 telocentric chromosomes, consisted of 45–46 chromosomes. The generation time of Ehrlich ascites tumor cells is 15–20 h; consequently, its "philoge- netic" age is $3 \cdot 10^4$ cell generations. During all this period the cells were reproduced asexually, i.e., they did not undergo meiosis. If the theoretical concepts, according to which stDNAs play a defined role in meiosis, are taken into consideration, it remains an enigma, why stDNA did not undergo any structural changes. The presented data demonstrate that stDNAs must play a defined role also in somatic cells.

According to Jones (1978), constitutive heterochromatin can also carry out so- matic functions, such as, for example, participation in formation of the chromocenter- nucleolus structure. Besides, it has long been known that heterochromatin affects the expression of the proximal genes (the position effect). Jones assumes that the position effect may contain the clue to the solution of stDNA significance.

In summarizing the considered material, we come to a conclusion that the data available at present does not exclude completely the possibility of stDNA participa- tion in the processes occurring in somatic cells.

## 10.3  Hypotheses on the Functional Role of stDNA

The first hypothesis on the nature of repeating DNAs was proposed by Britten and Kohne (1966). It was assumed that repeating sequences could be formed by saltatory replication of some genes or their parts. Subsequently, new, functionally significant DNA sequences could originate from them on the basis of various genetic processes (point mutations, insertions, deletions, translocations).

At the present when the structural features of stDNA are known, it is quite evident that the strutural gene of any protein can hardly be formed de novo from stDNA through genetic changes. As concerns the moderately repetitive sequences scattered throughout the genome and to which regulatory functions are ascribed, they were formed, most likely, by dispersion of satellite-like blocks in the genome by translocations and inversions.

Somewhat later Walker (1968) suggested that stDNAs could play a crucial role in the so-called chromosomal housekeeping, i.e., in determining the folding pathways and structural reorganization of chromosomes observed at different steps of the cell cycle. However, due to data on hybridization in situ published in 1969, demonstrating localization of stDNA in regions of constitutive heterochromatin, this idea was for some time in the background.

All the following studies, devoted to the functional role of stDNA proceeded from the heterochromatic nature of stDNAs. In general, hypotheses on the functional role of heterochromatin (and, consequently, of stDNA) can be divided into two groups; hypotheses according to which heterochromatin determines chromosome recognition in the process of meiotic pairing for the initial alignment of homologous chromosomes as well as aggregation of functionally related regions of nonhomologous chromosomes in the interphase nucleus are attributed to the first group; the second group includes hypotheses concerning the role of heterochromatin in regulating recombination frequencies during meiosis.

### 10.3.1  Recognition Hypothesis

The early concepts of Yunis and Yasmineh (1971) were based on recognition of chromosome heterochromatic regions. They assume that constitutive heterochromatin and, consequently, stDNAs, perform three important functions in a cell.

(1) Heterochromatin protects vitally important parts of the genome. From the authors viewpoint, heterochromatin forms a screen around the nucleolar organizer to shield rRNA cistrons against mutations and crossing-overs. The necessity of this protection is demonstrated by the conservatism of these cistrons. It is known also that plant and animal crossing-over in heterochromatin occurs with less frequency than in euchromatin.

(2) Hetrochromatin determines the aggregation both of homologous and nonhomologous chromosomes. One example is the prolonged aggregation of heterochromatin regions bearing nucleolar organizers during the mitotic and meiotic cycles. Furthermore, centromeric heterochromatin can determine the "centromeric strength" maintaining an exact segregation of chromosomes during cell division.

(3) The specificity of constitutive heterochromatin, determined by the set of stDNAs, may lead to an "incompatibility barrier" preventing pairing of heterochromatin of closely related species. Consequently, constitutive heterochromatin can provide evolutionary versatility and speciation.

In a review paper Walker (1971) supported the idea that the role of stDNAs consists in chromosome recognition and maintenance of the centromeric strength which determines the behavior of chromosomes in meiosis. He notes that it is not the centromere itself that determines the centromeric strength, but the heterochromatin adjoining the centromere.

The recognition hypothesis was developed further by Brutlag et al. (1977a) who showed that each chromosome of a *D.melanogaster* haploid set is characterized by a specific location of individual stDNAs in the centromeric heterochromatin. It was suggested that stDNA of homologous chromosomes determines meiotic pairing, while satellites in different chromosomes determine more general recognition processes leading to the formation of a chromocenter.

The hypothesis advanced by Corneo (1978) is based also on the recognition of chromosomes by stDNA-containing heterochromatin. He assumed that the appearance of stDNAs enhances formation of sexual reproduction in eukaryotes at the time when crossing-over between the two heterogametic sex chromosomes is suppressed, to keep the different genes of the sex chromosomes separate.

According to Corneo, this cold occur during the formation of any new stDNa or during a change of its quantity in one of the sex chromosomes. Subsequently, these properties could be passed on to nonsexual chromosomes by translocations in meiosis. When stDNAs in two different populations of one species become so different that homologous chromosome pairing is suppressed, the hybrids will become sterile. Thus, according to this hypothesis, stDNAs determine the mechanism of reproductive isolation and subsequent speciation.

A new hypothesis on the role of stDNA was suggested by Skinner et al. (1974). It was noted that the number of nucleotides in stDNA basic repeats is less than the minimum nucleotide number required to form a stable duplex (up to 1974 complex stDNAs were not yet discovered). The hypothesis assumed that destabilized satellites could pair with short homologous sequences of a more complex DNA and participate in chromosome recognition. Separate strands of guinea pig and other stDNAs were shown to form complexes with a number of double-helix prokaryotic and eukaryotic DNAs (Skinner 1977). However, there are no data at present which could confirm the existence of such a mechanism in vivo.

The possible functional role of satellite DNA has been considered also by other authors. Mazrimas and Hatch (1972) postulated that the presence of satellites provides a greater genetic flexibility to an organism. From a study of the quantitative content in kangaroo rats, the authors concluded that a higher content of stDNA is characteristic of more primitive species. It was, however, noted that such a correlation is not observed in crustaceans (Skinner et al. 1974).

## 10.3.2 Hypothesis of Regulation of Recombination

A completely different point of view has been developed by another group of authors. The first report in this direction was published by Miklos and Nankivell (1976). Their study of stDNAs and heterochromatin of three species of the *Atractomorpha* grass-hopper permitted them to abandon earlier postulated hypotheses on stDNA function. In their opinion, stDNAs can only regulate the frequency and position of recombinations. They do not deny a defined role of stDNAs in homologue recognition and in chromosome pairing, but for this to be done a small amount of DNA is sufficient, and it may be not observed as detectable heterochromatin. These conclusions are based on the following facts.

(1) There is no visible autosomal heterochromatin in *A.australis*, *A.species-1* contains a large centromeric block, while the third species, *A.similis*, contains both centro-meric and telomeric blocks.
(2) "Hidden" stDNAs localized in the heterochromatin of *A.similis* and *A.species-1* are absent in *A.australis*. Most of the postulated stDNA functions, namely, chromosome pairing, homologue recognition, centromeric strength, etc., are effi-ciently performed in *A.australis* which has neither stDNAs nor autosomal hetero-chromatin. The only difference between the compared species is the frequency and position of recombinations which correlate with the presence of heterochromatin.

A more generalized consideration of the above is discussed in a review paper by John and Miklos (1979). The authors categorically reject the idea of stDNA participa-tion in the processes of meiotic pairing and chromosome recognition. They report data on *D.melanogaster* heterochromatin studies demonstrating that X-chromosomes with as much as 80% of their heterochromatin deleted still segregate normally from each other. Furthermore, a considerable part of *D.melanogaster* X and Y chromo-somes in which stDNAs are located do not have pairing ability, and the pairing sites in the heterochromatin regions of sex chromosomes do not contain stDNAs.

There have been cases when some chromosomes, completely free of hetero-chromatin, paired normally and functioned with the same competence as hetero-chromatin-containing chromosomes.

According to John and Miklos, stDNAs (at least some of them) do not perform any somatic function, but play an essential role only in germ-line cells, affecting the process of recombination between homologous chromosomes during meiosis. The authors substantiate this conception by a large body of data confirming a correlation between the heterochromatin variability and recombination frequency. Figure 77 illustrates the effect of heterochromatin on euchromatin recombination in *D.melano-gaster*.

The fact that heterochromatin (and, consequently, stDNA) does not determine homologue recognition means that satellite DNAs do not affect fertiliy, in this case the differences in stDNA cannot serve as an incompatibility barrier.

It should be noted, however, that John and Miklos and other authors developing this approach do not exclude the possibility of some other functions of hetero-chromatin both in germ-line and somatic cells.

Bostock (1980) is in agreement that heterochromatin exerts a considerable effect on meiotic recombination in euchromatin segments of chromosomes (Fig. 78). He presumes that stDNA-containing constitutive heterochromatin affects the genetic constitution of the genome and, thus, is an object of selection. Selection does not occur for a defined stDNA sequence, but simply for the DNA structure contributing to the formation of a condensed heterochromatin state. The variability of the stDNA

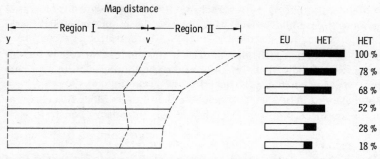

**Fig. 77.** Effect of X-heterochromatin reduction on the recombination of two euchromatic regions (y–v and v–f of the X-chromosome of *D.melanogaster*) (John and Miklos 1979)

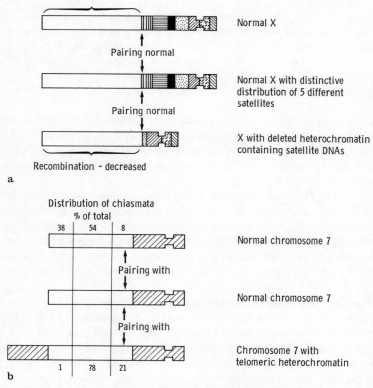

**Fig. 78.** Participation of satellite heterochromatin in the process of pairing and meiotic recombination (Bostock 1980). **a** The *D.melanogaster* X-chromosome containing five stDNAs in the region of centromeric heterochromatin displays normal pairing with the X-chromosome deleted of heterochromatin. However, recombination in the euchromatic region is reduced. The presence of telomeric heterochromatin inhibits recombination although pairing proceeds normally. **b** The presence of telomeric heterochromatin in *Atractomorpha similis* chromosome 7 depresses the recombination, although pairing occurs normally

quantity (and, consequently that of the heterochromatin) can provide a more rapid change of the genome than could be achieved only by the mutation of structural genes.

The precise mechanism of heterochromatin participation in the controlling recombination is still unknown. Usually, recombinations take place in the euchromatic part of the chromosome, in places often distal from stDNA-containing regions. It has been demonstrated that the addition of a stDNA-rich heterochromatin block to a chromosome leads to a change in chromosome length and this introduces constraints on recombination (Jones 1978).

"Is regulation of recombination a sufficient selective force to explain the maintenance and amplification of satellite DNA?" Walker (1978) contends that a final answer to this question cannot be given at present, although there are undeniable data evidencing that heterochromatin blocks modify the pattern of crossing-over in some species.

According to Walker, the chromosome structure itself requires a DNA in it with a mechanical and structural function; an hrDNA is necessary for chromosomes undergoing meiosis and mitoses. There must be many pathways by which these DNAs determine the structure of eukaryotic chromosomes: each species and even each individuum can have different optimal combinations to solve this problem.

### 10.3.3 Is stDNA Selfish?

The new theoretical concepts on the functional role of excess DNA in the cell have been developed in recent years. It has been long known and discussed in the literature that the amount of DNA in a chromosome haploid set of eukaryotes considerably exceeds that in structural and regulatory genes (Southern 1974). Thus, *Drosophila melanogaster* contains about 5000 genes in the genome; this amount corresponds to about $5 \cdot 10^6$ bp of a DNA. But total DNA of *D.melanogaster* contains $10^8$ bp, i.e., 20 times more of the gene material. The amount of DNA per haploid set of chromosomes in mammals is even higher, reaching, on the average, $3 \cdot 10^9$ bp. In some species of fish and amphibians the DNA amount is an order of magnitude higher. This paradox is explained by the fact that eukaryotes contain an excess of DNA, whose functions are not that of coding for proteins or regulating the expression of structural genes.

According to new viewpoints the excess DNA is unnecessary, incidental, with no phenotype expression whose only function is survival within genomes (Orgel and Crick 1980; Doolittle and Sapienza 1980). According to these authors, simple sequence and moderately repeated DNAs may act as selfish DNAs. Part of the unique DNAs, which do not code for any proteins and do not participate in the regulation of gene activity expression, are also attributed to selfish DNAs. Introns detected in eukaryotic genes are assigned also to this class of DNA.

The selfish DNA does not make any contribution to the phenotype. It only burdens energetically the cell that contains it. The appearance of selfish DNAs within the genome can be compared to the spread of a not too harmful parasite with its host. It is considered that a cell occasionally finds some use for some part of the selfish DNA, one of the possible applications being the control mechanisms of gene activity.

The authors of the hypotheses assume that amplification or deletion of selfish DNA is likely to occur through unequal crossing-over during meiosis or mitosis. The processes of formation and elimination of selfish DNA sequences must be balanced, though the situation can be in constant change.

It is difficult at present to conceive experimental approaches that could strictly prove or disprove this hypothesis. The assignment of satellite DNAs to the selfish class, useless to a certain degree, means the extension of analogous viewpoints also to constitutive heterochromatin as the main repository of stDNAs. Such a conception can hardly find many adherents.

### 10.3.4 Concluding Remarks

A characteristic detail is observed in the hypotheses on the functional role of stDNAs. These reports usually begin with an introductory phrase on the role of stDNAs, but further deal exclusively with questions related to the functional role of heterochromatin. Heterochromatin is a "macroscopic" structure, and, as yet, no stDNA structural data are needed to explain its functions. This again is evidence of the scarcity of data on heterochromatin. No hypothesis devoted to heterochromatin function requires the DNA contained in it to be satellite-like and not, for example, unique.

Two aspects should be clearly distinguished here: the first is the compactization of stDNA-containing chromatin and, probably, of its flanking regions, with the formation of a condensed state known as "constitutive heterochromatin"; the second is the function of the constitutive heterochromatin proper, a supramolecular structure, functioning as a qualitatively new unit.

What are the functions of stDNA itself? StDNA apparently participates in one way or another in the formation of constitutive heterochromatin. The functions of constitutive heterochromatin should be considered separately. It follows from the previous chapters that at present the question of heterochromatin function is more of a cytogenetic problem rather than a molecular-biological one.

The key to the solution of the role of stDNA is in providing answers to the following questions: Why is the presence of stDNA requisite for chromatin compactization? What structural stDNA features determine formation of the condensed state of chromatin? Why is stDNA not necessary to maintain a temporary condensed state of chromatin, a facultative heterochromatin?

# References

Ackerman EJ (1983) Molecular cloning and sequencing of OAX DNA: an abundant gene family transcribed and activated in *Xenopus* oocytes. EMBO J 2: 1417–1422

Altenburger W, Hörz W, Zachau HG (1977) Comparative analysis of three guinea pig satellite DNAs by restriction nucleases. Eur J Biochem 73: 393–400

Arrighi FE, Hsu TC (1971) Localization of heterochromatin in human chromosomes. Cytogenetics (Basel) 10: 81–86

Arrighi FE, Mandel M, Bergendahl J, Hsu TC (1970) Buoyant densities of DNA of mammals. Biochem Genet. 4: 367–376

Avila J, Montejo de Garcini E, Wandosell F, Villasante A, Sogo JM, Villanueva N (1983) Microtubule-associated protein $MAP_2$ preferentially binds to a dA/dT sequence present in mouse satellite DNA. EMBO J 2: 1229–1234

Barnes SR, Webb DA, Dover G (1978) The distribution of satellite and main-band DNA components in the *Melanogaster* species subgroup of *Drosophila*. I. Fractionation of DNA in actinomycin D and distamycin A density gradients. Chromosoma (Berl) 67: 341–363

Beauchamp RS, Mitchell AR, Buckland RA, Bostock CJ (1979) Specific arrangements of human satellite III DNA sequences in human chromosomes. Chromosoma (Berl) 71: 153–166

Beattie WG, Skinner DM (1972) The diversity of satellite DNAs of *Crustacea*. Biochim Biophys Acta 281: 169–178

Bedbrook JR, Jones J, O'Dell M, Thompson RD, Flavell RB (1980) A molecular description of telomeric heterochromatin in *Secale* species. Cell 19: 545–560

Bendich AJ, Anderson RS (1974) Novel properties of satellite DNA from muskmelon. Proc. Natl. Acad. Sci. USA 71: 1511–1515

Bendich AJ, Taylor WC (1977) Sequence arrangement in satellite DNA from the muskmelon. Plant Physiol. (Bethesda) 59: 604–609

Bendich AJ, Ward BL (1980) On the evolution and functional significance of DNA sequence organization in vascular plants. In: Leaver CJ (ed) Genome organization and expression in plants. Plenum, New York, pp 17–30

Beridze TG (1972) DNA nuclear satellites of the genus *Phaseolus*: Biochim Biophys Acta 262: 393–396

Beridze TG (1975) DNA nuclear satellites of the genus *Brassica*: variation between species. Biochim Biophys Acta 395: 274–279

Beridze TG (1980a) Satellite DNAs of higher plants. Ph D Thesis, AN Bakh Ins. Biochem. Acad. Sci. USSR, Moscow

Beridze TG (1980b) Properties of satellite DNAs of higher plants. Plant Sci Lett 19: 325–338

Beridze TG (1980c) The properties of satellite DNA of *Citrus limon*. Mol Biol (Mosc) 14: 126–135

Beridze TG, Bragvadze GP (1976) On the properties of satellite DNA of *Phaseolus vulgaris* Mol Biol (Mosc) 6: 1279–1289

Beridze TG, Odintsova MS, Sissakyan NM (1967) Distribution of DNA components of bean leaves in the fractions of cell structures. Mol Biol (Mosc) 1: 142–153

Bernstine EG (1978) Satellite DNA content of chromatin fractions isolated from *Eco*RI-digested mouse liver nuclei. Exp Cell Res 113: 205–208

Billings PC, Orf JW, Palmer DK, Talmage DA, Pan CG, Blumenfeld M (1979) Anomalous electrophoretic mobility of *Drosophila* phosphorylated H1 histone: is it related to the compaction of satellite DNA into heterochromatin? Nucleic Acids Res 6: 2152–2164

Birnstiel ML, Speiers J, Purdom I, Jones K, Loening UE (1968) Properties and composition of the isolated ribosomal DNA satellite of Xenopus laevis. Nature (Lond) 219: 454–463

Biro PA, Carr-Brown A, Southern EM, Walker PMB (1975) Partial sequence analysis of mouse satellite DNA: evidence for short range periodicities. J Mol Biol 94: 71–86

Blin N, Stephenson EC, Stafford DW (1976) Isolation and some properties of a mammalian ribosomal DNA. Chromosoma (Berl) 58: 41–50

Blumenfeld M (1979) Phosphorylated H1 histone in *Drosophila melanogaster*. Biochem Genet 17: 163–166

Blumenfeld M, Fox AS, Forrest HS (1973) A family of three related satellite DNAs in *Drosophila virilis*. Proc Natl Acad Sci USA 70: 2772–2775

Blumenfeld M, Orf JW, Sina BJ, Kreber RA, Callaham MA, Snyder LA (1978) Satellite DNA, H1 histone, and heterochromatin in *Drosophila virilis* Cold Spring Harbor Symp Quant Biol 40(1): 273–275

Böck H, Alber S, Zhang X-Y, Fritton H, Igo-Kemenes T (1984) Positioning of nucleosomes in satellite I containing chromatin of rat liver. J Mol Biol 176: 131–154

Bond HE, Flamm WG, Burr HE, Bond SB (1967) Mouse satellite DNA. Further studies on its biological and physical characteristics and its intracellular localization. J Mol Biol 27: 289–302

Bonner TI, Brenner DJ, Neufeld BR, Britten RJ (1973) Reduction in the rate DNA reassociation by sequence divergence. J Mol Biol 81: 123–135

Bonnewell V, Fowler RF, Skinner DM (1983) An inverted repeat borders a fivefold amplification in satellite DNA. Science (Wash DC) 221: 862–865

Borst P, Ruttenberg GJCM (1966) Renaturation of mitochondrial DNA. Biochim Biophys Acta 114: 645–647

Bostock CJ (1980) A function for satellite DNA. Trends Biochem Sci 5: 117–119

Bostock CJ, Christie S, Lauder IJ, Hatch FT, Mazrimas JA (1976) S phase patterns of replication of different satellite DNAs in three species of *Dipodomys* (kangaroo rat). J Mol Biol 108: 417–433

Bostock CJ, Gosden JR, Mitchell AR (1978) Localization of a male-specific DNA fragment to a sub-region of the human Y chromosome. Nature (Lond) 272: 324–328

Botchan MR (1974) Bovine satellite I DNA consists of repetitive units 1400 base pairs in length. Nature (Lond) 251: 288–292

Bragvadze GP (1983) Properties of satellite DNA of three plant species of subtribe *Citrinae*. Biokhimiya 48: 54–61

Bragvadze GP, Beridze TG (1976) Studies of the genome of some citric plants. In: Proc Georgian Conf Biochem Cultured Plants, Metsniereba, Tbilisi, pp 86–87

Bragvadze GP, Beridze TG (1983) Satellite DNAs of citric plants. Biokhimiya 48: 673–677

Britten RJ, Davidson EH (1976) Studies on nucleic acid reassociation kinetics: Empirical equations describing DNA reassociation. Proc Natl Acad Sci USA 73: 415–419

Britten RJ, Kohne DE (1966) Nucleotide sequence repetition in DNA. Carnegie Inst Wash Year Book 65: 78–106

Britten, RJ, Kohne DE (1968) Repeated sequences in DNA. Science (Wash DC) 161: 529–540

Britten RJ, Graham DE, Neufeld BR (1974) Analysis of repeating DNA sequences by reassociation. In: Grossman L, Moldave K (eds) Methods Enzymol vol 29(E) Academic, New York, pp 363–418

Brown SDM, Dover GA (1980) The specific organization of satellite DNA sequences on the X-chromosome of *Mus Musculus*: partial independence of chromosome evolution. Nucleic Acids Res 8: 781–792

Brown FL, Musich PR, Maio JJ (1979) The repetitive sequence structure of component α DNA and its relationship to the nucleosomes of the African green monkey. J Mol Biol 131: 777–799

Brutlag DL (1980) Molecular arrangement and evolution of heterochromatic DNA. Annu Rev Genet 14: 121–144

Brutlag D, Peacock WJ (1979) Sequences of the 1.672 g/cm³ satellite DNA of *Drosophila melanogaster*. J Mol Biol 135: 565–580

Brutlag D, Appels R, Dennis ES, Peacock WJ (1977a) Highly repeated DNA in *Drosophila melanogaster*. J Mol Biol 112: 31–47

Brutlag D, Fry K, Nelson T, Hung P (1977b) Synthesis of hybrid bacterial plasmids containing highly repeated satellite DNA. Cell 10: 509–519

Brutlag D, Carlson M, Fry K, Hsieh TS (1978) DNA sequence organization in *Drosophila* heterochromatin. Cold Spring Harbor Symp Quant Biol 40(1): 1137–1146

Bünemann H, Müller W (1978) Base specific fractionation of double stranded DNA: affinity chromatography on a novel type of adsorbant. Nucleic Acids Res 5: 1059–1074

Buongiorno-Nardelli M, Amaldi F (1970) Autoradiographic detection of molecular hybrids between rRNA and DNA in tissue sections. Nature (Lond) 225: 946–948

Calladine CR (1982) Mechanics of sequence-dependent stacking of bases in B-DNA. J Mol Biol 161: 343–352

Capesius I (1979) Isolation and characterization of native AT-rich satellite DNA from nuclei of the orchid *Cymbidium*. FEBS Lett 68: 255–258

Capesius I (1979) Isolation and characterization of satellite DNA from mustard seedlings. Plant Syst Evol 133: 1–13

Capesius I, Bierweiler B, Bachmann K, Rücker W, Nagl W (1975) An A+T-rich satellite DNA in a monocotyledonous plant, *Cymbidium*. Biochim Biophys Acta 395: 67–73

Carlson M, Brutlag D (1979) Different regions of a complex satellite DNA vary in size and sequence of the repeating unit. J Mol Biol 135: 483–500

Chambers CA, Schell MP, Skinner DM (1978) The primary sequence of a crustacean satellite DNA containing a family of repeats. Cell 13: 97–110

Chilton M-D (1973) Theoretical explanation of mouse satellite DNA renaturation kinetics. Nature New Biol 246: 16–17

Chilton M-D (1975) Ribosomal DNA in a nuclear satellite of tomato. Genetics 81: 469–483

Chuang CR, Saunders GF (1974) Complexity of human satellite A DNA. Biochem Biophys Res Commun 57: 1221–1230

Chun EHL, Vaughan MH, Rich A (1963) The isolation and characterization of DNA associated with chloroplast preparations. J Mol Biol 7: 130–141

Cooke HJ, Hindley J (1979) Cloning of human satellite III DNA: different components are on different chromosomes. Nucleic Acids Res 6: 3177–3197

Cordeiro M, Wheeler L, Lee CS, Kastritsis CD, Richardson RH (1975) Heterochromatic chromosomes and satellite DNAs of *Drosophila nasutoides*. Chromosoma (Berl) 51: 65–73

Cordeiro-Stone M, Lee CS (1976) Studies on the satellite DNAs of *Drosophila nasutoides*: Their buoyant densities, melting temperatures, reassociation rates and localizations in polytene chromosomes. J Mol Biol 104: 1–24

Corneo G (1978) Satellite DNAs in eukaryotes: a non-adaptive mechanism of speciation which originated with sexual reproduction? Experientia (Basel) 34/9: 1141–1142

Corneo G, Moore C, Sanadi DR, Grossman L, Marmur J (1966) Mitochondrial DNA in yeast and some mammalian species. Science (Wash DC) 151: 687–689

Corneo G, Ginelli E, Polli E (1967) A satellite DNA isolated from human tissues. J Mol Biol 23: 619–622

Corneo G, Ginelli E, Soave C, Bernardi G (1968) Isolation and characterization of mouse and guinea pig satellite deoxyribonucleic acids. Biochemistry 7: 4373–4379

Corneo G, Ginelli E, Polli E (1970a) Different satellite DNAs of guinea pig and ox. Biochemistry 9: 1565–1571

Corneo G, Ginelli E, Polli E (1970b) Repeated sequences in human DNA. J Mol Biol 48: 319–327

Corneo G, Ginelli E, Polli E (1971) Renaturation properties and localization in heterochromatin of human satellite DNAs. Biochim Biophys Acta 247: 528–534

Corneo G, Zardi L, Polli E (1972) Elution of human satellite DNAs on a methylated albumin kieselguhr chromatographic column: isolation of satellite DNA IV. Biochim Biophys Acta 269: 201–204

Corneo G, Meazza D, Bregni M, Tripputi P, Nelli LC (1982) Restriction enzyme studies on human highly repeated DNAs. Experimentia (Basel) 38: 454–457

Cortadas J, Macaya G, Bernardi G (1977) An analysis of the bovine genome by density gradient centrifugation: fractionation in $Cs_2SO_4$/3,6-Bis (acetatomercurimethyl) dioxane density gradient. Eur J Biochem 76: 13–19

Cramer JH, Farelly FW, Rownd RH (1976) Restriction endonuclease analysis of ribosomal DNA from *saccharomyces cerevisiae* Mol Gen Genet 148: 233–242

Darling SM, Grampton JM, Williamson R (1982) Organization of a family of highly repetitive sequences within the human genome. J Mol Biol 154: 51–63

Dennis ES, Gerlach WL, Peacock W J (1980) Identical polypyrimidine-polypurine satellite DNAs in wheat and barley. Heredity 44: 349–366

Deumling B (1981) Sequence arrangement of a highly methylated satellite DNA of a plant *Scilla*: a tandemly repeated inverted repeat. Proc Natl Acad Sci USA 78: 338–342

Deumling B, Nagl W (1978) DNA characterization, satellite DNA localization, and nuclear organization in *Tropaeolum majus*. Cytobiologie 16: 412–420

Deumling B, Sinclair J, Timmis JV, Ingle J (1976) Demonstration of satellite DNA components in several plant species with the $Ag^+ - Cs_2SO_4$ gradient technique. Cytobiologie 13: 224–232

Diaz MO, Barsacchi-Pilone G, Mahon KA, Gall JG (1981) Transcripts from both strands of a satellite DNA occur on lampbrush chromosome loops of the newt *Notophthalmus*. Cell 240: 649–659

Donehower L, Gillespie D (1979) Restriction site periodicities in highly repetitive DNA of primates. J Mol Biol 134: 805–834

Donehower L, Furlong C, Gillespie D, Kurnit D (1980) DNA sequence of baboon highly repeated DNA: evidence for evolution by nonrandom unequal crossovers. Proc Natl Acad Sci USA 77: 2129–2133

Doolittle WF, Sapienza C (1980) Selfish genes, the phenotype paradigm and genome evolution. Nature (Lond) 284: 601–603

Edelman M, Epstein HT, Schiff JA (1966) Isolation and characterization of DNA from the mitochondrial fraction of *Euglena*. J Mol Biol 17: 463–469

Emery HS, Weiner AM (1981) An irregular satellite sequence is found at the termini of the linear extrachromosomal rDNA in *Dictyostelium discoideum*. Cell 26: 411–419

Endow SA (1977) Analysis of *D.melanogaster* satellite IV with restriction endonuclease Mbo II. J Mol Biol 114: 441–449

Endow SA, Gall JG (1975) Differential replication of satellite DNA in polyploid tissues of *Drosophila virilis*. Chromosoma (Berl) 50: 175–192

Endow SA, Polan ML, Gall JG (1975) Satellite DNA sequences of *Drosophila melanogaster*. J Mol Biol 96: 665–692

Fittler F (1977) Analysis of the α-satellite DNA from African green monkey cells by restriction nucleases. Eur J Biochem 74: 343–352

Fittler F, Zachau HG (1979) Subunit struture of α-satellite DNA containing chromatin from African green monkey cells. Nucleic Acids Res 7: 1–13

Flamm WG, Bond HE, Burr HE (1966) Density-gradient centrifugation of DNA in a fixed-angle rotor. A higher order of resolution. Biochim Biophys Acta 129: 310–319

Flamm WG, Walker PMB, McCallum M (1969) Some properties of the single strands isolated from the DNA of the nuclear satellite of the mouse (*Mus musculus*). J Mol Biol 40: 423–443

Fodor I, Beridze T (1980a) Characterization of plant satellite DNA using restriction nucleases. Mol Biol Rep 6: 17–20

Fodor I, Beridze T (1980b) Structural organization of plant ribosomal DNA. Biochem Int 1: 493–501

Frenster JH, Allfrey VG, Mirsky AE (1963) Repressed and active chromatin isolated from interphase lymphocytes. Proc Natl Acad Sci USA 50: 1026–1029

Frommer M, Prosser J, Tkachuk D, Reisner AH, Vincent PC (1982) Simple repeated sequences in human satellite DNA. Nucleic Acids Res 10: 547–563

Fry K, Brutlag D (1979) Detection and resolution of closely related satellite DNA sequences by molecular cloning. J Mol Biol 135: 581–593

Fry K, Salser W (1977) Nucleotide sequences of HS-α satellite DNA from kangaroo rat *Dipodomys ordii* and characterization of similar sequences in other rodents. Cell 12: 1069–1084

Fry K, Poon R, Whitcome P, Idriss J, Salser W, Mazrimas J, Hatch F (1973) Nucleotide sequence of HS-β satellite DNA from kangaroo rat *Dipodomys ordii*. Proc Nat Acad Sci USA 70: 2642–2646

Gaillard C, Doly J, Cortadas J, Bernardi G (1981) The primary structure of bovine satellite 1.715. Nucleic Acids Res 9: 6069–6082

Gall JG (1974) Free ribosomal RNA genes in the macronucleus of *Tetrahymena*. Proc Nat Acad Sci USA 71: 3078–3081

Gall JG, Cohen EH, Polan ML (1971) Repetitive DNA sequences in *Drosophila*. Chromosoma (Berl) 33: 319–344

Gall JG, Cohen EH, Atherton DD (1973) The satellite DNAs of *Drosophila virilis*. Cold Spring Harbor Symp Quant Biol 38: 417–421

Gama-Sosa MA, Wang RY-H, Kuo KC, Gehrke CW, Ehrlich M (1983) The 5-methylcytosine content of highly repeated sequences in human DNA. Nucleic Acids Res 11: 3087–3095

Gauze GG (1977) Mitochondrial DNA. Nauka, Moscow (in Russian)

Gillis M, De Ley J, De Cleene M (1970) The determination of molecular weight of bacterial genome DNA from renaturation rates. Eur J Biochem 12: 143–153

Gosden JR, Mitchell AR, Buckland RA, Clayton RP, Evans HJ (1975) The location of four human satellite DNAs on human chromosomes. Exp Cell Res 92: 148–158

Gosden JR, Lawrie SS, Cooke HJ (1981) A cloned repeated DNA sequence in human chromosome heteromorphisms. Cytogenet Cell Genet 29: 32–39

Gottesfeld JM, Melton DA (1978) The length of nucleosome-associated DNA is the same in both transcribed and nontranscribed regions of chromatin. Nature (Lond) 273: 317–319

Graf H, Fittler F, Zachau HG (1979) Studies on the organization of the α-satellite DNA from African green monkey cells using restriction nucleases and molecular cloning. Gene (Amst) 5: 93–110

Gray DM, Skinner DM (1974) A circular dichroism study of the primary structures of three crab satellite DNA's rich in A:T base pairs. Biopolymers 13: 843–852

Green BR, Gordon MP (1966) Replication of choloroplast DNA of tobacco. Science (Wash DC) 152: 1071–1075

Grimaldi G, Singer MF (1983) Members of the *Kpn*I family of long interspersed repeated sequences join and interrupt α-satellite in the monkey genome. Nucleic Acids Res 11: 321–338

Grisvard J, Tuffet-Anghileri A (1980) Variations in the satellite DNA content of *Cucumis melo* in relation to dedifferentiation and hormone concentration. Nucleic Acids Res 8: 2843–2858

Gupta RC (1983) Nucleotide sequence of a reiterated rat DNA fragment. FEBS Lett 164: 175–180

Hagerman PJ (1984) Evidence for the existece of stable curvature of DNA in solution. Proc Natl Acad Sci USA 81: 4632–4636

Harbers K, Spencer JH (1978) Sequence studies on mouse L-cell satellite DNA by base-specific degradation with T4 endonuclease IV. Biochim Biophys Acta 520: 521–530

Harbers K, Harbers B, Spencer JH (1974) Nucleotide clusters in deoxyribonucleic acids. X. Sequences of the pyrimidine oligonucleotides of mouse L-cell satellite DNA. Biochem Biophys Res Commun 58: 814–821

Harel J, Hanania N, Tapiero H, Harel L (1968) RNA replication by nuclear satellite DNA in different mouse cells. Biochem Biophys Res Commun 33: 696–701

Hatch FT, Mazrimas JA (1970) Satellite DNA's in the kangaroo rat. Biochim Biophys Acta 224: 291–294

Hatch FT, Mazrimas JA (1974) Fractionation and characterization of satellite DNAs of the kangaroo rat (*Dipodomys ordii*). Nucleic Acids Res 1: 559–575

Hearst JE, Cech TR, Marx KA, Rosenfeld A, Allen JR (1974) Characterization of the rapidly renaturing sequences in the main CsCl density bands of *Drosophila*, mouse and human DNA. Cold Spring Harbour Symp Quant Biol 38: 329–339

Hemleben V, Grierson D, Dertmann H (1977) The use of equilibrium centrifugation in actinomycin-caesium chloride for the purification of ribosomal DNA. Plant Sci Lett 9: 129–135

Hemleben V, Leweke B, Roth A, Stadler J (1982) Organization of highly repetitive satellite DNA of two *Cucurbitaceae* species (*Cucumis melo* and *Cucumis sativus*). Nucleic Acids Res 10: 631–644

Hennig W (1972) Highly repetitive DNA sequences in the genome of *Drosophila hydei* II. Occurrence in polytene tissues. J Mol Biol 71: 419–431

Holmquist G (1975) Organization and evolution of *Drosophila virilis* heterochromatin. Nature (Lond) 257: 503–506

Horvath P, Hörz W (1981) The compaction of mouse heterochromatin as studied by nuclease digestion. FEBS Lett 134: 25–28

Hörz W, Altenburger W (1981) Nucleotide sequence of mouse satellite DNA. Nucleic Acids Res 9: 683–696

Hörz W, Zachau HG (1977) Characterization of distinct segments in mouse satellite DNA by restriction nucleases. Eur J Biochem 73: 383–392

Hörz W, Fittler F, Zachau HG (1983) Sequence specific cleavage of African green monkey α-satellite DNA my micrococcal nuclease. Nucleic Acids Res 11: 4275–4285

Hourcade D, Dressler D, Wolfson J (1973) The amplification of ribosomal RNA genes involves a rolling circle intermediate. Proc Natl Acad Sci USA 70: 2926–2930

Hsieh T-S, Brutlag D (1979a) Sequence and sequence variation within the 1.688 g/cm³ satellite DNA of *Drosophila melanogaster*. J Mol Biol 135: 465–481

Hsieh T-S, Brutlag DL (1979b) A protein that preferentially binds *Drosophila* satellite DNA. Proc Natl Acad Sci USA 76: 726–730

Hutton JR, Wetmur JG (1973) Length dependence of the kinetic complexity of mouse satellite DNA. Biochem Biophys Res Commun 52: 1148–1155

Igo-Kemenes T, Omori A, Zachau HG (1980) Non-random arrangement of nucleosomes in satellite I containing chromatin of rat liver. Nucleic Acids Res 8: 5377–5390

Ingle J, Pearson GG, Sinclair J (1973) Species distribution and properties of nuclear satellite DNA in higher plants. Nature New Biol 242: 193–197

Ingle J, Timmis JN, Sinclair J (1975) The relationship between satellite deoxiribonucleic acid, ribosomal ribonucleic acid gene redundancy, and genome size in plants. Plant Physiol (Bethesda) 55: 496–501

Ivanov IG, Markov GG (1980) Conservatism of mouse satellite DNA in transplantable tumors. In: Zadrazil S, Sponar J (eds) DNA – recombination, interaction and repair. Pergamon, Oxford pp 341–348

John HA, Birnstiel ML, Jones KW (1969) RNA-DNA hybrids at the cytological level. Nature (Lond) 223: 582–587

John B, Miklos GLG (1979) Functional aspects of satellite DNA and heterochromatin. Int Rev Cytol 58: 1–114

Jones KW (1970) Chromosomal and nuclear location of mouse satellite DNA in individual cells. Nature (Lond) 225: 912–915

Jones KW (1978) Speculations on the functions of satellite DNA in evolution. Z Morphol Anthropol 69: 143–171

Kavenoff R, Klotz LC, Zimm BH (1973) On the nature of chromosome-sized DNA molecules. Cold Spring Harbor Symp Quant Biol 38: 1–8

Kemp JD, Sutton DW (1976) A chemical and physical method for determining the complete base composition of plant DNA. Biochim Biophys Acta 425: 148–156

Kirk JTO (1971) Will the real chloroplast DNA please stand up? In: Boardman NK, Linnane AW, Smillie RM (eds) Autonomy and biogenesis of mitochondria and choloroplasts, North-Holland, Amsterdam, pp 267–276

Kit S (1961) Equilibrium sedimentation in density gradient of DNA preparations from animal tissues. J Mol Biol 3: 711–716

Kopecka H, Macaya G, Cortadas J, Thiery J-P, Bernardi G (1978) Restriction enzyme analysis of satellite DNA components from the bovine genome. Eur J Biochem 84: 189–195

Kornberg A, Bertsch LL, Jackson JF, Khorana HG (1964) Enzymatic synthesis of deoxyribonucleic acid. XVI. Oligonucleotides as templates and the mechanism of their replication. Proc Natl Acad Sci USA 51: 315–318

Kurnit DM (1979) Satellite DNA and heterochromatin variants: the case for unequal mitotic crossing over. Hum Genet 47: 169–186

Kurnit DM, Maio JJ (1974) Variable satellite DNA's in the African green monkey *Cercopithecus aethiops*. Chromosoma (Berl) 45: 387–400

Kurnit DM, Schildkraut CL, Maio JJ (1972) Single-strand interactions of mouse satellite DNA. Biochim Biophys Acta 259: 297–312

Kurnit DM, Shafit BR, Maio JJ (1973) Multiple satellite deoxyribonucleic acids in the calf and their relation to the sex chromosomes. J Mol Biol 81: 273–284

Laird C, McCarthy B (1968) Magnitude of interspecific nucleotide sequence variability in *Drosophila*. Genetics 60: 303–322

Lam BS, Carroll D (1983) Tandemly repeated DNA sequences from *Xenopus laevis*. I. Studies on sequence organization and variation in satellite 1 DNA (741 base-pair repeat). J Mol Biol 165: 567–585

LaMarca ME, Allison DP, Skinner DM (1981) Irreversible denaturation mapping of a pyrimidine-rich domain of a complex satellite DNA. J Biol Chem 256: 6475–6479

Lapeyre J-N, Beattie WG, Dugaiczyk A, Vizard D, Becker FF (1980) *Eco*RI-generated reiterated components of the rat genome. I. Sequence of two (92 and 93 bp) related DNA fragments. Gene (Amst) 10: 339–346

Lauer GD, Klotz LC (1975) Determination of the molecular weight of *Saccharomyces cerivisiae* nuclear DNA. J Mol Biol 95: 309–326

Lee TNH, Singer MF (1982) Structural organization of α-satellite DNA in a single monkey chromosome. J Mol Biol 161: 323–342

Levinger L, Varshavsky A (1982) Protein D1 preferentially binds A + T-rich DNA in vitro and is a component of *Drosophila melanogaster* nucleosomes containing A + T-rich satellite DNA. Proc Natl Acad Sci USA 79: 7152–7156

Lica L, Hamkalo B (1983) Preparation of centromeric heterochromatin by restriction endonuclease digestion of mouse L929 cells. Chromosoma (Berl) 88: 42–49

Luck DJL, Reich E (1964) DNA in mitochondria of *Neurospora crassa*. Proc Natl Acad Sci USA 52: 931–938

Macaya G, Cortadas J, Bernardi G (1978) An analysis of the bovine genome by density-gradient centrifugation. Preparation of the dG + dC-rich DNA components. Eur J Biochem 84: 179–188

Maio JJ, Brown FL, Musich PR (1977) Subunit structure of chromatin and the organization of eukaryotic highly repetitive DNA: recurrent periodicities and models for the evolutionary origins of repetitive DNA. J Mol Biol 117: 637–655

Manuelidis L (1977) A simplified method for preparation of mouse satellite DNA. Anal Biochem 78, 561–568

Manuelidis L (1978) Complex and simple sequences in human repeated DNAs. Chromosoma (Berl) 66: 1–21

Manuelidis L (1981) Consensus sequence of mouse satellite DNA indicates it is derived from 116 basepair repeats. FEBS Lett 129: 25–28

Manuelidis L, Wu JC (1978) Homology between human and simian repeated DNA. Nature (Lond) 276: 92–94

Maresca A, Singer MF (1983) Deca-satellite: a highly polymorphic satellite that joins α-satellite in the African green monkey genome. J Mol Biol 164: 493–511

Marini JC, Levene SD, Crothers DM, Englund PT (1982) Bent helical structure in kinetoplast DNA. Proc Natl Acad Sci USA 79: 7664–7668

Maroteaux L, Heilig R, Dupret D, Mandel JL (1983) Repetitive satellite-like sequences are present within or upstream from 3 avian protein-coding genes. Nucleic Acids Res 11: 1227–1243

Marx KA, Hearst JE (1975) The two component kinetic analysis method. Evidence for two renaturing components in mouse satellite DNA and *Dipodomys ordii* HS-β satellite DNA. J Mol Biol 98, 355–368

Marx KA, Allan JR, Hearst JE (1976) Chromosomal localization by in situ hybridization of the repetitious human DNA families and evidence of their satellite DNA equivalents. Chromosoma (Berl) 59: 23–42

Mathew ChGP, Goodwin GH, Igo-Kemenes T, Johns EW (1981) The protein composition of rat satellite chromatin. FEBS Lett 125: 25–29

Matsuda K, Siegel A (1967) Hybridization of plant ribosomal RNA to DNA: the isolation of a DNA component rich in ribosomal RNA cistrons. Proc Natl Acad Sci USA 58: 673–680

Mattoccia E, Comings DE (1971) Buoyant density and satellite composition of DNA of mouse heterochromatin. Nature New Biol 229: 175–176

Mazin AL, Muchovatova LM, Shuppe NG, Vanyushin BF (1984) In DNA of *Drosophila melanogaster* and *Drosophila virilis* 5-methylcytosine is absent. Dokl Akad Nauk SSSR 276: 760–762

Mazrimas JA, Hatch FT (1972) A possible relationship between satellite DNA and the evolution of kangaroo rat species (genus *Dipodomys*). Nature New Biol 240: 102–105

Mazrimas JA, Hatch FT (1977) Similarity of satellite DNA properties in the order *Rodentia*. Nucleic Acids Res 4: 3215–3227

Mazrimas JA, Balhorn R, Hatch FT (1979) Separation of satellite DNA chromatin and main band DNA chromatin from mouse brain. Nucleic Acids Res 7: 935–946

McCutchan T, Hsu H, Thayer RE, Singer MF (1982) Organization of African green monkey DNA at junctions between α-satellite and other DNA sequences. J Mol Biol 157: 195–211

Melli M, Ginelli E, Corneo G, DiLernia R (1975) Clustering of the DNA sequences complementary to repetitive nuclear RNA of HeLa cells. J Mol Biol 93: 23–38

Meselson M, Stahl FW, Vinograd J (1957) Equilibrium sedimentation of macromolecules in density gradients. Proc Natl Acad Sci USA 43: 581–588

Meyer MW, Lippincott JA (1967) Deoxyribonucleic acids of normal and crown-gall tumor tissues of primary "Pinto" bean leaves. Plant Physiol (Bethesda) 42: 553

Meyerhof W, Tappeser B, Korge E, Knöchel W (1983) Satellite DNA from *Xenopus laevis*: comparative analysis of 745 and 1037 base pair *Hind*III tandem repeats. Nucleic Acids Res 11: 6997–7009

Miklos GLG, Gill AC (1981) The DNA sequences of cloned complex satellite DNAs from Hawaiian *Drosophila* and their bearing on satellite DNA sequence conservation. Chromosoma (Berl) 82: 409–427

Miklos GLG, John B (1979) Heterochromatin and satellite DNA in man: properties and prospects. Am J Hum Genet 31: 264–280

Miklos GLG, Nankivell RN (1976) Telomeric satellite DNA functions in regulating recombination. Chromosoma (Berl) 56: 143–167

Mitchell AR, Beauchamp RS, Bostock CJ (1979) A study of sequence homologies in four satellite DNAs of man. J Mol Biol 135: 127–149

Moritz KB, Roth GE (1976) Complexity of germline and somatic DNA in *Ascaris*. Nature (Lond) 259: 55–57

Müller F, Walker P, Aeby P, Neuhaus H, Felder H, Back E, Tobler H (1982) Nucleotide sequence of satellite DNA contained in the eliminated genome of *Ascaris lumbricoides*. Nucleic Acids Res 10: 7493–7510

Mullins JI, Blumenfeld M (1979) Satellite Ic: a possible link between the satellite DNA of *D.virilis* and *D.melanogaster*. Cell 17: 615–621

Musich PR, Brown FL, Maio JJ (1977a) Subunit structure of chromatin and the organization of eucaryotic highly repetitive DNA: nucleosomal proteins associated with a highly repetitive mammalian DNA. Proc Natl Acad Sci USA 74: 3297–3301

Musich PR, Maio JJ, Brown FL (1977b) Subunit structure of chromatin and the organization of eucaryotic highly repetitive DNA: indications of a phase relation between restriction sites and chromatin subunits in African green monkey and calf nuclei. J Mol Biol 117: 657–677

Musich PR, Brown FL, Maio JJ (1978) Mammalian repetitive DNA and the subunit structure of chromatin. Cold Spring Harbor Symp Quant Biol 42(2): 1147–1160

Musich PR, Brown FL, Maio JJ (1980) Highly repetitive component α and related alphoid DNAs in man and monkey. Chromosoma (Berl) 80: 331–348

Musich PR, Brown FL, Maio JJ (1982) Nucleosome phasing and micrococcal nuclease cleavage of African green monkey component α DNA. Proc Natl Acad Sci USA 79: 118–122

Nandi US, Wang JC, Davidson N (1965) Separation of deoxyribonucleic acids by Hg(II) binding and Cs$_2$SO$_4$ density gradient centrifugation. Biochemistry 4: 1687–1696

Omori A, Igo-Kemenes T, Zachau HG (1980) Different repeat lengths in rat satellite I DNA containing chromatin and bulk chromatin. Nucleic Acids Res 8: 5363–5375

Orgel LE, Crick FHC (1980) Selfish DNA: the ultimate parasite. Nature (Lond) 284: 604–607

Pages MJM, Roizes GP (1984) Nature and organization of the sequence variations in the long-range periodicity calf satellite DNA I. J Mol Biol 173: 143–157

Pardue ML (1975) Repeated DNA sequences in the chromosomes of higher organisms. Genetics 79: 159–170

Pardue ML, Gall JG (1969) Formation and detection of RNA-DNA hybrid molecules in cytological preparations. Proc Natl Acad Sci USA 63: 378–383

Pardue ML, Gall JG (1970) Chromosomal localization of mouse satellite DNA. Science (Wash DC) 168: 1356–1358

Pardue ML, Gall JG (1972) Molecular cytogenetics. In: Sussman M (ed) Molecular genetics and developmental biology. Prentice Hall Inc Englewood Cliffs, pp 65–99

Pashev IG, Markov GG (1978) Hydroxyapatite thermal chromatography of chromatin, elution pattern of mouse satellite DNA in partially dehistonized and reconstituted chromatin. Int J Biochem 9: 307–312

Pashev IG, Dimitrov SI, Ivanov IG, Markov GG (1983) Histone acetylation in chromatin containing mouse satellite DNA. Eur J Biochem 133: 379–382

Patterson JB, Stafford DW (1971) Characterization of sea urchin ribosomal satellite deoxyribonucleic acid. Biochemistry 10: 2775–2779

Peacock WJ, Brutlag D, Goldring E, Appels R, Hinton CW, Lindsley DL (1974) The organization of highly repeated DNA sequences in *Drosophila melanogaster* chromosomes. Cold Spring Harbor Symp Quant Biol 38: 405–416

Peacock WJ, Lohe AR, Gerlach WL, Dunsmuir P, Dennis ES, Appels R (1978) Fine structure and evolution of DNA in heterochromatin. Cold Spring Harbor Symp Quant Biol 42(1): 1121–1135

Peacock WJ, Dennis ES, Rhoades MM, Pryor AJ (1981) Highly repeated DNA sequence limited to knob heterochromatin in maize. Proc Natl Acad Sci USA 78: 4490–4494

Pech M, Igo-Kemenes T, Zachau HG (1979a) Nucleotide sequence of a highly repetitive component of rat DNA. Nucleic Acids Res 7: 417–432

Pech M, Streeck RE, Zachau HG (1979b) Patchwork structure of a bovine satellite DNA. Cell 18: 883–893

Pietras DF, Bennett KL, Siracusa LD, Woodworth-Gutai M, Chapman VM, Gross KW, Kane-Haas C, Hastie ND (1983) Construction of a small *Mus musculus* repetitive DNA library: identification of a new satellite sequence in *Mus musculus*. Nucleic Acids Res 11: 6969–6983

Plucienniczak A, Skowronsky J, Jaworski J (1982) Nucleotide sequence of bovine 1.715 satellite DNA and its relation to other bovine satellite sequences. J Mol Biol 158: 293–304

Pöschl E, Streeck RE (1980) Prototype sequence of bovine 1.720 satellite DNA. J Mol Biol 143: 147–153

Polli E, Ginelli E, Bianchi P, Corneo G (1966) Renaturation of calf thymus satellite DNA. J Mol Biol 17: 305–308

Prescott DM, Murti KG, Bostock CJ (1973) Genetic apparatus of *Stylonychia* sp. Nature (Lond) 242: 576–600

Rabinowitz M, Sinclair J, De Salle L, Haselkorn R, Swift H (1965) Isolation of deoxyribonucleic acid from mitochrondria of chick embryo heart and liver. Proc Natl Acad Sci USA 53: 1126–1133

Rae P (1970) Chromosomal distribution of rapidly reannealing DNA in *Drosophila melanogaster*. Proc Natl Acad Sci USA 67: 1018–1025

Rae PMM, Barnett TR, Babbit DG (1976) Factors influencing the yield of satellite DNA in extractions from *Drosophila virilis* and *Drosophila melanogaster* adults and embryos. Biochim Biophys Acta 432: 154–160

Ranjekar PK, Pallotta D, Lafontaine JG (1978) Analysis of plant genomes. III. Denaturation and reassociation properties of cryptic satellite DNAs in barley (*Hordeum vulgare*) and wheat (*Triticum aestivum*). Biochim Biophys Acta 520: 103–110

Renkawitz R (1979) Isolation of twelve satellite DNAs from *Drosophila hydei*. Macromolecules 1: 133–136

Renkawitz-Pohl R, Kunz W (1975) Underreplication of satellite DNAs in polyploid ovarian tissue of *Drosophila virilis*. Chromosoma (Berl) 49: 375–382

Roizes G (1976) A possible structure for calf satellite DNA I. Nucleic Acids Res 3: 2677–2696

Roizes GP, Pages M (1982) Conserved and divergent sequences of bovine satellite DNAs. In: Dover GA, Flavell RB (eds) Genome evolution. Academic, London, pp 95–111

Roizes G, Pages M, Lecou C (1980) The organization of the long range periodicity calf satellite DNA I variants as revealed by restriction enzyme analysis. Nucleic Acids Res 8: 3779–3792

Rosenberg H, Singer M, Rosenberg M (1978) Highly reiterated sequences of simiansimiansi-miansimiansimian. Science (Wash DC) 200: 394–402

Roth GE (1979) Satellite DNA properties of the germ line limited DNA and the organization of the somatic genomes in the nematodes *Ascaris suum* and *Parascraris equorum*. Chromosoma (Berl) 74: 355–371

Rubin CM, Deininger PL, Houck CM, Schmid CW (1980) A dimer satellite sequence in bonnet monkey DNA consists of distinct monomer subunits. J Mol Biol 136: 151–167

Salomon R, Kaye AM, Herzberg M (1969) Mouse nuclear satellite DNA: 5-methylcytosine content, pyrimidine isoplith distribution and electron microscopic appearance. J Mol Biol 43: 581–592

Salser W, Bowen S, Browne D, El Adli F, Fedoroff N, Fry K, Heindell H, Paddock G, Poon R, Wallace B, Whitcome P (1976) Investigation of the organization of mammalian chromosomes at the DNA sequence level. In: Genome organization in Higher Organisms. Fed Proc 35: 23–35

Sano H, Sager R (1982) Tissue specificity and clustering of methylated cytosines in bovine satellite I DNA. Proc Natl Acad Sci USA 79: 3584–3588

Saunders GF, Hsu TC, Getz MJ, Simes EL, Arrighi FE (1972) Locations of a human satellite DNA in human chromosomes. Nature New Biol 236: 244–246

Schildkraut CL, Maio JJ (1968) Studies on the intranuclear distribution and properties of mouse satellite DNA. Biochim Biophys Acta 161: 76–93

Schildkraut CL, Marmur J, Doty P (1962) Determination of the base composition of deoxyribonucleic acid from its buoyant density in CsCl. J Mol Biol 4: 430–443

Schweber MS (1974) The satellite bands of the DNA of *Drosophila virilis*. Chromosoma (Berl) 44: 371–382

Sealy L, Hartley J, Donelson J, Chalkley R, Hutchison N, Hamkalo B (1981) Characterization of a highly repetitive sequence DNA family in rat. J Mol Biol 145: 291–318

Selsing E, Leslie AGW, Arnott S, Gall JG, Skinner DM, Southern EM, Spencer JH, Harbers K (1976) Conformations of satellite DNAs. Nucleic Acids Res 3: 2451–2457

Seshadri M, Ranjekar PK (1979) Genome characterization of three plant species belonging to the genus *Phaseolus*. Indian J Biochem Biophys 16: 1–5

Shen C-KJ, Hearst JE (1977) Detection of long-range sequence order in *Drosophila melanogaster* satellite DNA IV by a photochemical crosslinking reaction and denaturation microscopy. J Mol Biol 112: 495–507

Shen C-KJ, Wiesehahn G, Hearst JE (1976) Cleavage patterns of *Drosophila melanogaster* satellite DNA by restriction enzymes. Nucleic Acids Res 3: 931–951

Shipp WS, Kieras FJ, Haselkorn R (1965) DNA associated with tobacco chloroplasts. Proc Natl Acad Sci USA 54: 207–213

Shmookler Reis RJ, Biro PA (1978) Sequence and evolution of mouse satellite DNA. J Mol Biol 121: 357–374

Shmookler Reis R, Timmis JN, Ingle J (1981) Divergence, differential methylation and interspersion of melon satellite DNA sequences. Biochem J 195: 723–734

Sinclair J, Wells R, Deumling B, Ingle J (1975) The complexity of satellite deoxyribonucleic acid in a higher plant. Biochem J 149: 31–38

Singer DS (1979) Arrangement of highly repeated DNA sequence in the genome and chromatin of the African green monkey. J Biol Chem 254: 5506–5514

Singer MF (1982) Highly repeated sequences in mammalian genomes. Int Rev Cytol 76: 67–112

Sissoëff I, Grisvard J, Guillé E (1976) Studies on metal ions – DNA interactions: specific behaviour of reiterative DNA. Prog Biophys Mol Biol 31: 165–199

Skinner DM (1967) Satellite DNAs in the crabs *Gecarcinus lateralis* and *Cancer pagurus*. Proc Natl Acad Sci USA 58: 103–110

Skinner DM (1977) Satellite DNAs. Bioscience 27: 790–796

Skinner DM, Beattie WG (1973) $Cs_2SO_4$ gradients containing both $Hg^{++}$ and $Ag^+$ effect the complete separation of satellite deoxyribonucleic acids having identical densities in neutral CsCl gradients. Proc Natl Acad Sci USA 70: 3108–3110

Skinner DM, Beattie WG (1974) Characterization of a pair of isopycnic twin crustacean satellite deoxyribonucleic acids, one of which lacks one base in each strand. Biochemistry 13: 3922–3929

Skinner DM, Triplett LL (1967) The selective loss of DNA satellites on deproteinization with phenol. Biochem Biophys Res Commun 28: 892–897

Skinner DM, Beattie WG, Kerr MS, Graham DE (1970) Satellite DNAs in *Crustacea*: two different components with the same density in neutral CsCl gradients. Nature (Lond) 227: 837–839

Skinner DM, Beattie WG, Blattner FR, Stark BP, Dahlberg JE (1974) The repeat sequence of a hermit crab satellite deoxyribonucleic acid is $(-T-A-G-G-)_n\cdot(-A-T-C-C-)_n$. Biochemistry 13: 3930–3937

Skinner DM, Bonnewell V, Fowler RF (1983a) Sites of divergence in the sequence of a complex satellite DNA and several cloned variants. Cold Spring Harbor Symp Quant Biol 47: 1151–1157

Skinner DM, Fowler RF, Bonnewell V (1983b) Domains in simple sequences or alternating purines and pyrimidines are sites of divergence in a complex satellite DNA. In: Mechanisms of DNA replication and recombination. Liss, New York, pp 849–861

Sloof P, Bos JL, Konings FJM, Menke HH, Borst P, Gutteridge WE, Leon W (1983) Characterization of satellite DNA in *Trypanosoma brucei* and *Trypanosoma cruzi*. J Mol Biol 167: 1–21

Smith GP (1974) Unequal crossover and the evolution of multigene families. Cold Spring Harbor Symp Quant Biol 38: 507–513

Smith GP (1976) Evolution of repeated DNA sequences by unequal crossover. Science (Wash DC) 191: 528–535

Smith GP (1978) What is the origin and evolution of repetitive DNAs? Trends Biochem Sci 3: 34–36

Southern EM (1970) Base sequence and evolution of guinea-pig α-satellite DNA. Nature (Lond) 227: 794–798

Southern EM (1971) Effects of sequence divergence on the reassociation properties of repetitive DNAs. Nature New Biol 232: 82–83

Southern E (1974) Eukaryotic DNA. In: Burton K (ed) MTP Int Rev Sci Biochem Nucleic Acids 6: 101–139

Southern EM (1975) Long range periodicities in mouse satellite DNA. J Mol Biol 94: 51–69

Stambrook PJ (1981) Interspersion of mouse satellite deoxyribonucleic acid sequences. Biochemistry 20: 4393–4398

Steinemann M (1976) The in situ formation of DNA·DNA duplexes of *Drosophila virilis* satellite DNA. Chromosoma (Berl) 54: 339–348

Strauss F, Varshavsky A (1984) A protein binds to a satellite DNA repeat at three specific sites that would be brought into mutual proximity by DNA folding in the nucleosome. Cell 37: 889–901

Streeck RE (1981) Inserted sequences in bovine satellite DNA's. Science (Wash DC) 213: 443–445

Streeck RE (1982) A multicopy insertion sequence in the bovine genome with structural homology to the long terminal repeats of retroviruses. Nature (Lond) 298: 767–769

Streeck RE, Moritz KB, Beer K (1982) Chromatin diminution in *Ascaris suum*: nucleotide sequence of the eliminated satellite DNA. Nucleic Acids Res 10: 3495–3502

Sturm KS, Taylor JH (1981) Distribution of 5-methylcytosine in the DNA of somatic and germ-line cells from bovine tissues. Nucleic Acids Res 9: 4537–4546

Sueoka N (1961) Variation and heterogeneity of base compositions of deoxyribonucleic acids: a compilation of old and new data. J Mol Biol 3: 31–40

Sueoka N, Cheng T (1961) Natural occurrence of a DNA resembling the deoxyadenylate-deoxythymidilate polymer. Proc Natl Acad Sci USA 48: 1851–1856

Sutton WD, McCallum M (1971) Mismatching and reassociation rate of mouse satellite DNA. Nature New Biol 232: 83–85

Swartz MN, Trautner TA, Kornberg A (1962) Enzymatic synthesis of deoxyribonucleic acid. XI. Further studies on nearest neighbor base sequences in deoxyribonucleic acids. J Biol Chem 237: 1961–1967

Swift H (1965) Nucleic acids of mitochondria and chloroplasts. Am Nat 99: 201–227

Szybalski W (1968) Equilibrium ultracentrifugation in the caesium sulphate density gradient. In: Grossman L, Moldave K (eds) Methods Enzymol., Vol 12(B). Academic, New York, XII, pp 330–360

Takhtajan AL (1966) System and phylogeny of flowering plants (in Russian). Nauka Moscow

Taparowsky EJ, Gerbi SA (1982a) Sequence analysis of bovine satellite I DNA (1.715 g/cm³). Nucleic Acids Res 10: 1271–1281

Taparowsky EJ, Gerbi SA (1982b) Structure of 1.711 g/cm³ bovine satellite DNA: evolutionary relationship to satellite I. Nucleic Acids Res 10: 5503–5515

Tewari KK, Wildman SG (1966) Chloroplast DNA from tobacco leaves. Science (Wash DC) 153: 1269–1271

Tewari KK, Jayaraman J, Mahler HR (1965) Separation and characterization of mitochondrial DNA from yeast. Biochem Biophys Res Commun 21: 141–148

Thayer RE, Singer MF (1983) Interruption of an α-satellite array by a short member of the *Kpn*I family of interspersed highly repeated monkey DNA sequences. Mol Cell Biol 3: 967–973

Thayer RE, Singer MF, McCutchan TF (1981) Sequence relationships between single repeat units of highly reiterated African green monkey DNA. Nucleic Acids Res 9: 169–181

Timmis JN, Ingle J (1977) Variation in satellite DNA from some higher plants. Biochem Genet 15: 1159–1173

Travaglini EC, Petrovic J, Schultz J (1968) Two satellite cytoplasmic DNAs in *Drosophila*. J. Cell Biol 39: 136a

Travaglini EC, Petrovic J, Schultz J (1972) Satellite DNAs in the embryos of various species of the genus *Drosophila*. Genetics 72: 431–439

Ulyanov NB, Zhurkin VB (1984) Anisotropic flexibility of DNA depends on nucleotide sequence. Conformational calculations of tetrameric duplexes AAAA:TTTT, (AATT)$_2$, (TTAA)$_2$, GGGG:CCCC, (GGCC)$_2$ and (CCGG)$_2$. Mol Biol (Mosc) 18: 1664–1685

Varley JM, MacGregor HC, Erba HP (1980) Satellite DNA is transcribed on lampbrush chromosomes. Nature (Lond) 283: 686–688

Varshavsky A, Levinger L, Sundin O, Barsoum J, Ozkaynak E, Swerdlow P, Finley D (1983) Cellular and SV40 chromatin: replication, segregation, ubiquitination, nuclease-hypersensitive sites, HMG-containing nucleosomes, and heterochromatin-specific protein. Cold Spring Harbor Symp Quant Biol 47(1): 511–528

Vinograd J, Hearst JE (1962) Equilibrium sedimentation of macromolecules and viruses in a density gradient. Fortschr Chemie Org Naturst 20: 372–422

Wagner I, Capesius I (1981) Determination of 5-methylcytosine from plant DNA by high-performance liquid chromatography. Biochim Biophys Acta 654: 52–56

Waldfogel FA, Swartz MN (1971) Subcellular location of crab poly (dA − dT). Biochim Biophys Acta 246: 403–411

Walker PMB (1968) How different are the DNAs from related animals? Nature (Lond) 219: 228–232

Walker PMB (1971) "Repetitive" DNA in higher organisms. In: Butler JAV, Noble D (eds) Prog Biophys Mol Biol, Vol 23. Pergamon, Oxford, pp 147–190

Walker PMB (1978) Distribution of DNA base sequences. Introductory remarks: DNA and genes. Philos Trans R Soc Lond Biol Sci 283: 305–307

Wall LVM, Bryant JA (1981) Isolation and preliminary characterization of cryptic satellite DNA in pea. Phytochemistry (Oxf) 20: 1767–1771

Waring M, Britten RJ (1966) Nucleotide sequence repetition: A rapidly reassociating fraction of mouse DNA. Science (Wash DC) 154: 791–794

Weber JL, Cole RD (1982a) Chromatin fragments containing bovine 1.715 g · ml$^{-1}$ satellite DNA. Purification by chromatography on malachite green DNA affinity resin. J Biol Chem 257: 11774–11783

Weber JL, Cole RD (1982b) Chromatin fragments containing bovine 1.715 g · ml$^{-1}$ satellite DNA. Nucleosome structure and protein composition. J Biol Chem 257: 11784–11790

Wells RD, Büchi H, Kössel H, Ohtsuka E, Khorana HG (1967) Studies on polynucleotides. LXX. Synthetic deoxyribopolynucleotides as templates for the DNA polymerase of *Escherichia coli*: DNA-like polymers containing repeating tetranucleotide sequences. J Mol Biol 27: 265–272

Wetmur JG, Davidson N (1968) Kinetics of renaturation of DNA. J Mol Biol 31: 349–370

White R, Pasztor LM, Hu F (1975) Mouse satellite DNA in noncentromeric heterochromatin of cultured cells. Chromosoma (Berl) 50: 275–282

Wishe G, Corces VG, Avila J (1978) Differential binding of hog brain microtubule-associated proteins to mouse satellite versus bulk DNA preparations. Nature (Lond) 273: 403–405

Witney FR, Furano AV (1983) The independent evolution of two closely related satellite DNA elements in rats (*Rattus*). Nucleic Acids Res 11: 291–304

Wollenzien P, Barsanti P, Hearst JE (1977) Location and underreplication of satellite DNA in *Drosophila melanogaster*. Genetics 87, 51–65

Wolstenholme DR, Gross NJ (1968) The form and size of mitochondrial DNA of the red bean, *Phaseolus vulgaris*. Proc Natl Acad Sci USA 61: 245–252

Wu HM, Crothers DM (1984) The locus of sequence-directed and protein-induced DNA bending. Nature (Lond) 308: 509–513

Wu JC, Manuelidis L (1980) Sequence definition and organization of a human repeated DNA. J Mol Biol 142: 363–386

Wu KC, Strauss F, Varshavsky A (1983) Nucleosome arrangement in green monkey α-satellite chromatin. Superimposition of non-random and apparently random patterns. J Mol Biol 170: 93–117

Yunis JJ, Yasmineh WG (1970) Satellite DNA in constitutive heterochromatin of the guinea pig. Science (Wash DC) 168: 263–265

Yunis JJ, Yasmineh WG (1971) Heterochromatin, satellite DNA, and cell function. Science (Wash DC) 174: 1200–1209

Zardi L, Siri A, Santi L, Rovera G (1977) Heterogeneity of mouse satellite DNA on silver-cesium sulphate density gradient. FEBS Lett 79: 188–190

Zhang XY, Hörz W (1982) Analysis of highly purified satellite DNA containing chromatin from the mouse. Nucleic Acids Res 10: 1481–1494

Zhang XY, Hörz W (1984) Nucleosomes are positioned on mouse satellite DNA in multiple highly specific frames that are correlated with a diverged subrepeat of nine base-pairs. J Mol Biol 176: 105–129

Zhang XY, Fittler F, Hörz W (1983) Eight different highly specific nucleosome phases on α-satellite DNA in the African green monkey. Nucleic Acids Res 11: 4287–4306

Zhukowsky PM (1971) Cultured plants and their relatives (in Russian). Kolos, Leningrad

# Subject Index